The Body Within

Brill's Studies in Intellectual History

VOLUME 176

Brill's Studies on Art, Art History, and Intellectual History

VOLUME 3

The Body Within

Art, Medicine and Visualization

Edited by

Renée van de Vall
Robert Zwijnenberg

BRILL

LEIDEN • BOSTON
2009

On the cover: **Cogito Ergo Sum 1 Susan Aldworth 2002** mixed media 43 x 35 cms.

This book is printed on acid-free paper.

Library of Congress Cataloging-in-Publication Data

The body within : art, medicine and visualization / edited by Renée van de Vall and Robert Zwijnenberg.
p. cm. — (Brill's studies in intellectual history, 0920-8607 ; v. 176) (Brill's studies on art, art history, and intellectual history ; v. 3)
Includes bibliographical references and index.
ISBN 978-90-04-17621-8 (hbk. : alk. paper) 1. Human body. 2. Medicine and art. 3. Image analysis. I. Vall, Renée van de, 1956– II. Zwijnenberg, Robert, 1954– III. Title. IV. Series.

HM636.B656 2009
306.4'61—dc22

2009025163

ISSN 0920-8607
ISBN 978 9004 17621 8

PRINTED IN THE NETHERLANDS

CONTENTS

LIST OF ILLUSTRATIONS

1. *Leonardo and Female Interiority*

2. *Animals Inside: Anatomy, Interiority and Virtue in the Early Modern Dutch Republic*

3. *Depicting Skin: Microscopy and the Visual Articulation of Skin Interior 1820–1850*

4. *The Mind at Work: The Visual Representation of Cerebral Processes*

7. *Looking for a Sponge: How a Body Learns to be Affected by Ultrasound*

11. *Framing Interiority: Portraits in the Age of Genomics*

LIST OF CONTRIBUTORS

José van Dijck is a Professor of Media and Culture at the University of Amsterdam where she is currently the Dean of Humanities. Her research areas include media and science, (digital) media technologies, popularisation of science and medicine, and television and culture. Her latest book *Mediated Memories in the Digital Age*, in which she theorizes the relationship between media technologies and personal cultural memory, was published by Stanford University Press (2007).

Michael Hagner is Professor of Science Studies at the ETH Zürich. His current research projects include the history of cybernetics and the function of visualization in the sciences. His most recent publication is (ed. with E. Hörl), *Die Transformation des Humanen. Beiträge zur Geschichte der Kybenetik*. Frankfurt/Main 2008.

Mieneke te Hennepe is curator medical history at Museum Boerhaave: National Museum of the History of Science and Medicine in Leiden, the Netherlands. Her PhD thesis *Depicting Skin: Visual culture in nineteenth-century medicine* was awarded with a Research Prize of the Praemium Erasmianum Foundation in 2008. She specialises in visual representation and body history.

Rina Knoeff is postdoctoral research fellow in the Faculty of Humanities at the University of Leiden. She has worked on the history of medicine, chemistry and anatomy in relation to (natural) philosophy, theology and the arts. Currently she is director of a research programme entitled *Cultures of Collecting. The Leiden Anatomical Collections in Context*. She is the author of *Herman Boerhaave (1668–1738). Calvinist Chemist and Physician* (Amsterdam: Edita, 2002).

Babette Müller-Rockstroh is a Research Fellow at the Max-Planck-Institute for Social Anthropology in Halle, Germany. Her field of expertise lies in forging the disciplinary intersection of Science, Society &Technology Studies and Critical Medical Anthropology in the field of International Reproductive Health. Her most recent article, 'Ultraschall in Tansania. Ethnographische Explorationen in Medizintechnologietran sfer', will appear in Hadolt, B.; Hörbst, V. & Dilger, H.J. (eds). *Medizin im Kontext* (Frankfurt: Peter Lang, forthcoming 2009).

MAUD RADSTAKE received her MA degree in medical anthropology at the University of Amsterdam and her PhD at the Faculty of Arts and Culture of Maastricht University in the field of Technology and Society Studies. Since 2006, she works as a postdoctoral researcher and staff member at the Centre for Society and Genomics, based at the Radboud University in Nijmegen. In 2007 she published *Visions of illness. An endography of real-time medical imaging* (Delft: Eburon).

MIRIAM VAN RIJSINGEN is an art historian at the University of Amsterdam and co-director of The Arts & Genomics Center, University of Leiden. Her research includes the fields of art & science, 18th century to contemporary art, representational theory and spectatorship, and theories of the body and embodiment. Her most recent publication is "Prolific Encounters—Towards a Philosophy of Mutability", *Mutamorphosis: Challenging Arts and Sciences*, On-Line Conference Proceedings, CIANT, 9 pp. (forthcoming Spring 2009).

JENNY SLATMAN is assistant professor in the philosophy department at Tilburg University and researcher in the department of health, ethics and society at Maastricht University. Her research includes phenomenological analyses of embodied self-experience and bodily integrity. In 2008, she has published a monograph (in Dutch): *Strange Body: On Medical Interventions and Personal Identity.*

RENÉE VAN DE VALL is associate professor Media Culture at the Faculty of Arts and Social Sciences, Maastricht University. Her research focuses on philosophy of art and aesthetics, specifically on contemporary visual art. In 2008 she has published *At the Edges of Vision. A Phenomenological Aesthetics of Contemporary Spectatorship* (Ashgate).

GAIL WEISS is Professor of Philosophy and Human Sciences at The George Washington University (USA). Her research lies at the intersections of phenomenology and existentialism, feminist theory, and philosophy of literature. Her most recent monograph is *Refiguring the Ordinary* (Indiana University Press, 2008).

Robert Zwijnenberg is professor History and Theory of Contemporary Art in Relation to the Sciences and Technology at the Faculty of Humanities, Leiden University. He has published on aesthetics and on art-science interaction, including the preface for Ingeborg Reichle's, *Art in the Age of Technoscience: Genetic Engineering, Robotics, and Artificial Life in Contemporary Art* (Wien, Springer, 2009).

ACKNOWLEDGEMENTS

This volume would not have been possible without the generous support by a number of Dutch research institutions. First of all we would like to thank the Netherlands Organization for Scientific Research (NWO) that has financed the interdisciplinary research project *The Mediated Body* and the Faculty of Arts and Social Sciences of Maastricht University that has hosted it. Both institutions have also supported the project's international expert meetings, together with the Royal Netherlands Academy of Arts and Sciences (KNAW) and the Netherlands Graduate Research School of Science, Technology and Modern Culture (WTMC). Several of the chapters of this volume were initially presented as papers at the conference *The Interior Mediated Body*, organized at Maastricht University on January 30 en 31 2004. We thank the participants of this conference for their comments. Chapter 2, "Animals inside: anatomy, interiority and virtue in the early modern Dutch Republic" was first published in *Medizinhistorisches Journal* 43 (2008) pp. 1–19. The publisher of *MhJ* has kindly given permission for the article to be reproduced. Chapter 9, "Mediated Memories as Amalgamations of Mind, Matter, and Culture" was first published, in slightly different form, as a chapter in José van Dijck, *Mediated Memories in the Digital Age* (Stanford University Press, 2007). Stanford University Press has kindly given permission for the chapter to be reproduced.

INTRODUCTION

THE BODY WITHIN: ART, MEDICINE AND VISUALIZATION

Renée van de Vall

Just an occasional medium range shot in the popular hospital series *Grey's Anatomy*. Doctors and nurses are assembled around a patient, a pregnant woman who has lost an arm in a car accident. The woman in her bed forms the centre of the image. We have a full view of what's left of her arm, the cut forming a red spot against the white sheets. On the left hand side of the bed the plastic surgeon Sloan, nicknamed McSteamy, cautiously touches the arm stump with his gloved hand. On the right hand side the series' heroine, the intern Meredith Grey, is seen from behind moving a sensor over the woman's belly while gazing at the monitor of an ultrasound machine. Asks McSteamy: "Do we have an arm?" "Not yet", replies Meredith. The camera zooms in on the woman's face. "The baby has no father. He already lacks two arms. I really need my arm." A close up of McSteamy's face, on which we see compassion emerging. "Get me that arm," he commands, full of determination.

The scene is quite characteristic for the series' over the top condensation of contrasting emotions, like the crude professionalism of Sloan and the heartbreaking appeal of the woman. On a less obvious level, however, it also alludes to contrasting notions of embodiment. The arm that is missing in Sloan's question is a different arm than the one the woman is referring to and both have little in common with the gaping wound we are facing on the screen. Likewise, the immediate view on the body's interior the wound allows us has a completely different impact than the vague white spots figuring on the monitor. The cruel sight of the bloody stump is softened, however, and its morbidity eased through the careful touch of the surgeon's hand, a touch that is repeated, though more mechanically and clinically, in that of the ultrasound sensor Meredith is wielding.

This present day *Anatomy Lesson* nicely juxtaposes touch and sight, immediate and technically mediated perception, compassion and distance, and, not in the least, plays with the ambivalent feelings of fascination and disgust the body's interior usually evokes. The first glance

Meredith throws at the cut is tinged with horror and as a spectator one can hardly bear the sight of the wound without feeling an unpleasant tingle at a spot somewhere halfway the upper arm. In spite of the marvellous achievements of present day medicine and the ubiquity of images of the body's insides, gazing into the body isn't altogether a piece of cake yet, a cultural fact *Grey's Anatomy* cleverly plays upon. Moreover, the shot ingenuously portrays the body's multiplicity. What might appear to be a single entity, the woman's body, might be a 'multiple' (Mol 2002), as it is articulated in contrasting ways through different interactions. Heavily emotionally charged on the left hand side, in the direct juxtaposition of wounded flesh with careful touch, it becomes more thing-like on the right hand side, in spite of the fact that it is a pregnant belly that is being scanned. What the inner body is to us largely depends on what takes place outside its boundaries.

Paradoxes of interiority

It is these ambiguous connotations that make the cultural imagination of the body's interior throughout history such an intriguing subject for philosophers, historians, anthropologists and literary scholars alike. Close and remote at the same time, the body within could become the source of amazement at God's or Nature's intricate design but also a focus for feelings of anxiety, fear or disgust. Intimately connected with 'oneself' (however questionable such a notion may be, historically speaking) and hidden from view simultaneously, a possible locus of sensations of well-being but also of unease or pain, the body's inner structures and workings might be 'absent' most of the time (Leder 1999), yet as soon as they claim attention, can hardly leave one indifferent.

The inner body is both familiar and strange (Slatman 2008). Although it is central for our access to the world, our own access to it is highly limited. In contrast to the body's surface powers of perception, motility and expression, its internal capacities and processes are largely unavailable to conscious awareness and command, as Drew Leder has emphasized in his phenomenological study of bodily interiority. Compared with the precision, subtlety and diversity of exteroperception, interoperception is qualitatively poor, ambiguous with regard to location and discontinuous in space and time. Yet when the body's interior makes itself felt, we cannot remain dispassionate.

> The limited interoceptive vocabulary largely centers around sensations that are affectively charged. [... V]isceral sensations grip me from within, often exerting an emotional insistence. [...I]t is the discomforting or painful sensations that speak up most clearly: the crampy stomach, the heartburn, the insistent need for defecation. (Leder 1999, 40)

Although there have been changes in their phenomenological constitution, such experiential ambiguities seem to persist across historical epochs and cultures. Historian Jonathan Sawday has pointed to the divergent senses of corporeality speaking from late twentieth-century compared to pre-modern descriptions of bodily interiority, yet also notes significant continuities. One of these is the tension between the interior body's centrality to, and its inaccessibility for experience.

> Whatever process is at work, the sense of interiority is inescapably central to the experience of the body within history. Yet, a feature of our sense of interiority is that it can never be experienced other than at second-hand. We may look into other bodies, but very rarely are we allowed to pry into our own. We may become familiar with the generalized topography of the body, via different media—photographs, X-rays, illustration, anatomical illustration, written description, TV documentary—but al these 'voyages within' [...] are journeys of exploration which encounter bodies other than our own. They are passages into THE body, but not MY body. (Sawday 1995, 7)

The very few reports of occasions in which someone is confronted with a look into MY body, Sawday observes, are fraught with distress. Throughout history, a strong taboo surrounds the observation of one's own interior. Although the fascination and horror evoked by the sight of one's body within may be caught in variously framed vocabularies, in each case the experience is described as being profoundly disturbing.

Sawday's examples, however, mainly concern occasions in which patients are confronted with a sight on their interior through open wounds. We may ask, therefore, whether present day medical visualisation techniques mark a significant shift in the experience of bodily interiority. Starting in 1895 with the invention of X-ray technology and more specifically in the last decades of the 20th century with the introduction of ultrasound, endoscopy, CT, MRI and PET-scans, these new visualisation techniques enable medical researchers and practitioners to look inside living bodies without literally opening them. Moreover, with the possibility of printing medical images or displaying them on screens, the interior body has become available not only for the gaze of medical professionals but for that of its inhabitant as well. More and more often, patients are allowed to watch their own insides during diagnostic or surgical operations and their testimonies, although often still ambiguous, do not suggest the same deep disturbance as Sawday noted. Thus, the new experiential possibility opened by these modern techniques may have profound implications for the ways in which the relations between 'body', 'self', and 'world' are configured, both on the level of cultural discourses and practices and on the level of individual experiences. Even if we agree with Leder (as for instance Jenny Slatman's chapter in this volume) that the possibility of visually observing one's own colon does not resolve its experiential strangeness (Leder 1999, 44), the very fact of this comparatively direct and yet comparatively dispassionate observation forms a new type of phenomenological situation of which the meaning has hardly been explored.

All the same, these occasions of clinical observation will not happen very often in one's lifetime and they will be limited to specific organs or body parts. In general, bodily interiority will remain characterised by absence and will have to be made deliberately present to be experientially available. Against this background of inaccessibility, the means through which interior body parts are made available will largely shape how they are conceived. (e.g. Radstake 2007, 94–97) Precisely because most of the practices currently constituting medical visualisation are still rather exceptional and not yet completely taken for granted, the new techniques provide a challenging opportunity for scholarly reflection. As such, the investigation of their dynamics shows 'embodiment in the making', demonstrating to what extent the presencing of bodily interiority results from an interaction of many external agents, including other bodies, images, technological artefacts and organs. Furthermore, this investigation compels us to rethink the history of the body—or rather, the histories of bodies—and of bodily mediations from a new angle. Transitional moments in the constitution of embodiment such as we are currently witnessing prompt the question of how our conceptions of bodily interiority came into being in the first place.

The seemingly paradoxical result of this volume's focus on interiority is the dissolution of boundaries between 'within' and 'without', 'inside' and 'outside', between the body and its environment. The conclusion emerging from these chapters is that the body can neither be considered as a fixed, self-contained, self-sufficient entity, nor as clearly demarcated from the outside physical and social world. What a body is and what it is capable of, where it begins and ends, how it is cared for and used, how it is understood and experienced, has been subject to changes due to developments apparently 'outside' its boundaries, yet profoundly altering what it—at least in its own perception—contained and excluded. Moreover, to the question whether and how bodies are—scientifically, artistically, discursively, institutionally—*mediated*, the new medical visualization techniques add the question whether and how they are *mediatized*.[1] In addition to the discursive principles guiding practices of observation, registration, interpretation and treatment, the distinctive role of the media employed in these practices becomes more obvious and informs, in retrospect, our understanding of earlier visual articulations of the body's insides. Anatomical theatres, medical atlases and ultrasound machines have in common that they are not only tools for investigation or instruction, but serve as representation and communication media as well and can be analysed as such.

Bodily mediation as mediatization

Foregrounding the role of media in the mediation of bodily interiority means shifting attention from the content to the carrier of a representation. 'Medium' can be defined as:

> [a]ny object or device used for communicating a message by moving physical information over distance or preserving it through time. The medium links the sender to the receiver. A medium may be as simple as a carved stone or as complex as a satellite-linked television system. (De Fleur and Dennis 1998, 570)

This definition foregrounds the materials and technologies with which, and the forms in which, messages are created, transmitted and displayed. But there is more to a medium than the object or device that carries the message. Media are also defined by what has been called 'apparatus', that is, the particular arrangement of devices producing a specific viewing (or listening) experience and configuring the psychological disposition

[1] Several authors, among which Lisa Cartwright (1995) and José van Dijck (2005), have emphasized the close intertwinement of the development of medical visualisation technologies with that of media technologies. In contrast to van Dijck, however, this introduction uses the term 'mediatization', as a specific form of the more general term 'mediation'. A microscope, for instance, allows a technologically mediated form of perception; it is not a medium however, until the microscopic image is communicated to an audience larger than the microscopist alone.

of the spectator (Sturken and Cartwright 2001, 349)[2] and by 'protocols', that is the normative rules, habits and standards governing practices of production and use (Gitelman 2006, 5–8).[3] The chapters of this volume make clear that by foregrounding the media of visualisation, in their dimensions of technological devices, apparatuses and protocols, it becomes possible to articulate aspects of bodily mediation that an exclusive emphasis on the content of representations and discourses tends to overlook.

Take for instance the endoscope, an instrument for the inspection of the inner cavities of the body. In its simplest forms it consists of a tube or cable, a light source, lenses and a viewer. It can be inserted though the natural openings of the body, like the vagina or the rectum, or through a small incision. Since it has become possible to attach a very small photo or video camera, it does not only function as a perceptual but also as a representational and communication technology, providing stills or live video images on a television screen. What the camera sees is made visible for more than one spectator, often also for the patient, and made part of a communicative situation: the endoscope begins to function as (part of) a medium. When endoscopic images are being transported outside the clinical setting, they become part of cultural imagination. As José van Dijck has written, the development of endoscopy is 'inextricably bound up with innovations in media technologies, such as photography, video and computers.' (van Dijck 2005, 65) Vice versa, endoscopy 'profoundly affects the ways in which the interior body is conceptualized and represented in popular media such as television.' In particular, the dissemination of endoscopic imagery has popularised what van Dijck calls 'the endoscopic gaze', the surgeon's view not from the outside into the interior, but *from within.* (Ibid., 66)

[2] The theoretical concept 'cinematic apparatus' emerged in French film theory in the early 70's and is particularly associated with the work of Jean-Louis Baudry. In general, it designates 'the cultural determinations and effectivities of the cinematic machinery and of the disposition of production and exhibition technologies and techniques' with regard to the construction and positioning of the subject and its spectatorship (Rosen, 1986, p. ix, p. 282).

[3] Protocols may consist of social, economical and material relationships. 'So telephony includes the salutation 'Hello?'..., the monthly billing cycle, and the wires and cables that materially connect our phones. E-mail includes all of the elaborately layered technical protocols and interconnected service providers that constitute the Internet, but it also includes both the QWERTY keyboards on which email gets 'typed' and the shared sense people have of what the e-mail genre is.' (Gitelman 2006, 7–8)

The intriguing question not yet answered in this cultural analysis of endoscopic representation is whether and how this 'view from within' is reconnected with the individual patient's experience of his or her *own* body, which as far as visual experience is concerned, is always necessarily limited to a view from outside. Maud Radstake's 'endographic' study of medical imaging events argues that such incorporation is not self-evident but achieved (or not) in a concerted effort of patients, doctors, images and medical instruments. (chapter 7, see also Radstake 2007) The technical possibilities and constraints of medical visualisation techniques—in Radstake's chapter in this volume the example is ultrasound of the kidneys—do play a role in this effort, but their dimensions as media apparatus and protocol are just as important.

In the case of Radstake's studies the images were real-time, therefore existing while they were made, coinciding with the event of the imaging procedure. This assured the tactile and kinaesthetic immediacy of the connection between body, imaging instrument and image, but also the image's ephemeral presence. Photographic images would have had a different spatiotemporal relation to the imaged body: although more stable as artefacts, they would have lacked the mobile appearance of a subjective gaze (see for instance Sobchack 2004) and be severed from the tactile sensations produced by the physical connection of body, instrument and image. In addition, Radstake shows whether and how the images are experienced by the patients also depend on the communicative interaction between doctors and patients. The dialogues Radstake has recorded might be called protocols-in-emergence, clusters of codes, habits and conventions regulating the proprieties of a new communicative situation of a medium in development. Together, ultrasound's apparatus and protocols form what Radstake calls 'an entire network of visualisation... formed by physicians, patients and technological devices before, during and after an imaging event.' Whether or not ultrasound imaging alienates the patient from his or her body is not determined by the technology in itself, but results from the particular configuration of the entire network in which the agency of the patient is a constituent factor. In this conclusion, Radstake differs from critical judgements like those of Lisa Cartwright, according to whom the use of cinema in the sciences was primarily—though not exclusively—a strategy of control and domination (Cartwright 1995, xv).

Analysing medical visualisation technologies in terms of media apparatus and protocol proves particularly fruitful where technologies travel. Babette Mueller-Rockstroh's study of the use of the 'Western' technology of ultrasound in Northwest Tanzania demonstrates how thoroughly the 'message' the ultrasound image is supposed to convey results from the specific configurations of the ultrasound device's use by doctors, nurses and patients and the conflicting cultural discourses which meet in the imaging procedure. As Mueller-Rockstroh remarks, in the manufacturing hall in the Netherlands no one would have suspected that ultrasound machines would be able to depict jealousy or ancestral spirits. Transported into Tanzanian hospitals, ultrasound pregnancy scanning practice meets with women's conceptions and enactments of pregnancy mediated by local health care practices. Contrary to what one might expect, ultrasound scanning does not supplant but interacts with or even reinforces traditional notions of forces and entities inhabiting women's bodies. What ultrasound machines can see and show and under what conditions is an effect not only of the properties of the imaging device, but of the mutual shaping of the imaging technology and the social group of pregnant women, the hospital's protocols and, at a further remove, policies of the state and international organisations promoting the use of the new technology.

Shifting boundaries between within and without

I have taken these two chapters as points of departure because they clearly exemplify the main theses that frame this volume. Their investigations of individual cases of medical visualisation practices foreground the interconnectedness of inside and outside on the level of material practice and embodied experience and the role of visualisation techniques as media. To understand how inside and outside connect, one should investigate how the technology that makes the interior externally present figures as a medium, that is as a device for communication, within a specific spatiotemporal arrangement and as part of specific communicative procedures. Radstake's chapter describes the imaging procedure as an event, in which body parts that are usually absent from conscious experience are being made present simultaneously with the production of the images. The imaging event is both an exteriorization, the coming forward of a usually hidden part of one's insides as an object of knowledge and experience, and an interiorization, a transformation of subjective agency and experience. This making present is far more complicated than conventional notions of the relation between representation and experience suggest. The same technological device may partake in a variety of embodiments. According to the particular material and perceptual configurations through which patient bodies are connected to images, Radstake distinguishes no less than four different modes of attachment and four kinds of image-bodies.

This connectedness and multiplicity highlights the precarious status of what we might call the 'modern body' with its clear separation between inside and outside. Thus, Mueller-Rockstroh's descriptions of the processes occurring within the bodies of pregnant women in Tanzania—spirits entering and children disappearing in the stomach, foetuses being turned to the back—elicit the same kind of wonder as Barbara Duden has expressed in *Geschichte unter dem Haut*, a fascinating study of the notebooks of Dr. Johannes Storch, a German physician from around 1730. (Duden 1987) Whereas the modern body is conceived as an isolated and closed whole, of which the interior consists of functionally related organs arranged in a three-dimensional space, the bodies of Storch's eighteenth century female patients were on the one hand more open to influences of the outside world, but on the other hand not spatially continuous with that world. Relatively unaffected by the anatomical depictions of theoretical science, the inner body was not thought of in geometrical terms in medical practice and popular imagination. It was an opaque locus for hidden transformations that could only be read through outward signs. Bodily processes were primarily interpreted in terms of substances entering, travelling through and leaving the body, changing in form, colour and substance yet indistinct with regard to place and trajectory. The skin, rather than closing the body from the outside, was the surface on which interior metamorphoses disclosed themselves. Bodily excretions were scrutinised for the same reason, as indicative of the interior movements and changes of fluids.

For a different type of interior body to emerge, it would have to be made externally present and in this presentation visual media such as drawings have always played a pivotal role. Leonardo da Vinci's anatomical drawings were probably the first pictorial efforts to depict the body's interior in a more than schematic way, as a three dimensional space. This was not an easy achievement, Robert Zwijnenberg argues in his chapter on "Leonardo and Female Interiority". Zwijnenberg's chapter focuses on Leonardo's 1515 drawing of a foetus in the uterus. The way Leonardo drew the foetus, in a stooped position with its hands before the eyes, almost too large for the womb that holds it, is highly charged with emotion. According to Zwijnenberg, anatomical drawing confronted Leonardo with the pictorial paradox that the body's inner, the closed anatomical space, can only be depicted as an outside, as a surface. It is through charging this surface with emotions, the movements of the soul, which are 'inner' in another, metaphorical sense, that Leonardo solved this paradox. And emotional the inner body was for Leonardo. Comparing the drawing with Leonardo's 1483–9 version of *The Virgin of the Rocks*, Zwijnenberg shows that both share the mixed feelings of desire and fear for the female womb as origin of life on the one hand and dark and threatening place at the same time. For Leonardo, this fear and desire were closely related to the fear and desire of the anatomist to open the human body and by that very act destroying the order it aims to reveal.

In her chapter "Depicting Skin: Microscopy and the Visual Articulation of Skin Interior 1820–1850", Mieneke te Hennepe argues that a crucial factor in the passage from the 'open', pre-modern body to the 'closed' modern body is to be found in a new visual articulation of the skin emerging in the practice of microscopic depiction in Germany and France between 1820 and 1850. From an open porous cover, serving as a passageway for bodily fluids and substances, the skin became a functionally active, thick organ with its own inner structures. The emergence of this boundary organ signalled an important step towards the body as a closed-off entity with an internal milieu. Te Hennepe traces this development by showing how the combined practice of microscopic investigation and drawing articulated a skin that became visually 'thicker' as it came to harbour more and more inner structures. This depiction resulted in a standardised schematic picture used in anatomical works, books on skin diseases and popular encyclopaedias from the middle of the nineteenth century onwards.

Yet the schemata of scientific representation are but one way of presenting and it is questionable whether the 'modern', anatomical conception of the body within ever succeeded in completely excluding other conceptions. What Barbara Duden has noted about 18th century Germany and Mueller-Rockstroh makes clear for 21st century Tanzania may be applicable for other times and places as well, including our own: the possibility of the co-existence of several, often contrasting notions and practices concerning the body. In making these alternative interiors externally present, other kinds of entities than pictures or texts may serve as media as well. Rina Knoeff's chapter "Animals Inside. Anatomy, Interiority and Virtue in the Early Modern Dutch Republic" demonstrates the visual and tangible character of early modern conceptions of the body's insides by the story of Grietje Willems, who was reported to have thrown up a little dog. Grietje's story is but one of several eighteenth-century accounts of animals such as snakes, toads and slugs living inside the body and being vomited up. These creatures were more than mere imaginations, Knoeff argues; they were taken seriously and keenly investigated by contemporary anatomists and physiologists like Ruysch and literally and tangibly illustrate how people experienced and imagined the working of the interior body. The invention of the microscope and the ensuing discovery of all kinds of tiny animals living in the water reinforced the view that animals could end up inside human beings. Comparable to the pile worms undermining Dutch dykes and houses, these internal animals materially symbolised the pain of disease and the dangers of moral corruption.

Interiorities and identities

The dominance of the modern concept of the body as a self-contained entity is closely connected with notions of individuality and identity as being somehow located inside our bodies, the skin forming the boundary between 'self' and 'other'. (Slatman 2008, 61) Brain imagery is a particular illuminative topic with regard to this equation of identity with interiority, because of the intimate association of the mental functions of the brain with human subjectivity in modern Western culture. Michael Hagner's chapter "The Mind at Work. The Visual Representation of Cerebral Processes" explores the consequences of increasingly sophisticated imagery of brain activities for the anthropological understanding of human nature and identity. In contrast to older visualisation technologies, in new techniques such as fMRT (functional magnetic resonance tomography) the distinction between morphological and functional representation of the brain is dissolved. The image on the screen directly informs us about the activities the brain is involved in, supposedly constituting an open window onto psychological phenomena. The possibility to visualise increasingly specific mental activities could result in a substitution of the superficial insights of brain images for the in-depth probing of introspective and subjective conceptions of mental life, reducing its diversity and relevance.

Whereas Michael Hagner focuses on brain-images' adequacy or inadequacy as representations of mental life and fears for the reductive effect they could have in the world of everyday experience, Renée van de Vall's chapter starts from the opposite end. She asks under what conditions brain images might acquire such a role in everyday experience and how this role could be investigated. Drawing on the phenomenological concept of the body-image and on ethnographical studies of medical practice in the tradition of Actor-Network-Theory, she argues that it is crucial to investigate how people 'learn to be affected' by these images on the level of individual experience. Whether brain images reduce or enrich people's notions of themselves and others cannot be predicted beforehand, but depends on the way the images function in actual practice in the performative articulation of their subjective embodiment.

In Van de Vall's chapter, the distinction between subject and object and the equation of subjectivity with the mind and objectivity with the body, c.q. the brain, is questioned. Patients suffering from depression use brain images to separate their brains as a kind of unruly subjects from their 'real' selves. The next chapter by Jenny Slatman elaborates another kind of differentiation. "Transparent Bodies: Revealing the Myth of Interiority" deconstructs the notion of interiority as the locus of subjectivity by focusing on the phenomenological distinction between *Körper*, the body as mere object, and *Leib*, the body-subject that senses itself sensing. Slatman argues that interiority not only has a spatial, but also a psychological meaning, the inner and private experience or feeling of oneself. This experience is based on something exterior and spatial, however, the body encountered in its irreducible thinghood. This *Körperlichkeit* can be defined as an irreducible strangeness. Imaging technologies visualizing the body's interior intensify and externalize the distinction between *Leib* and *Körper* and, rather than unveiling the living body's interiority, picture its death.

The modern cultural association of individual identity and subjectivity with interiority is further questioned in the last three chapters. In "Mediated Memories as Amalgamations of Mind, Matter, and Culture", José van Dijck starts from the movie *Eternal Sunshine of the Spotless Mind* to discuss the question where autobiographical memory is located and if it can actually be erased from our lives. Autobiographical memory seems to be located in the biological brain, while on the other hand memories are derived from mediated objects, such as photos, home videos, or diaries. Personal cultural memory, however, is neither located *within* the brain nor *outside* in culture, but is the result of a complex interaction between brain, material objects, and the cultural matrix from which they arise. Digitization apparently transforms the scientific representation of our memories, exemplified by the use of fMRI scans, but at the same time digital technologies shape our cultural products of remembering. It seems unlikely that this transformation would leave former notions of personal cultural memory intact.

Digitized memory, according to van Dijck, like ordinary memory, is neither located inside the body nor outside it, but should be understood as an embodied experience mediated through the distributed agency of mind, computer, objects and technology. In a comparable manner, Gail Weiss challenges the idea of bodily autonomy as the marker of individual identity by foregrounding the body's relations to other bodies. In her chapter "Intertwined Identities" she argues that the complex ties that bind together conjoined twins reveal the intercorporeal connections that all human beings sustain with one another. Extensive media attention to operations that separate conjoined twins tends to overlook that those operations are in tension with a primary concern for the health and well-being of patients themselves. It is taken for granted that the conjoined twins' existence is impoverished and tragic and their physical separation a matter of the highest medical urgency. This assumption is motivated by a cultural anxiety regarding the patients' seemingly ambiguous identity, as the physical boundaries of the human body have always served to distinguish one person from another. Conjoined twins incarnate and generate fears of our intercorporeality.

Finally, the interlocking of the themes of the mediatized externalisation of bodily interiority and the dissolution of the boundaries between the body within and without comes to a close in Miriam van Rijsingen's chapter. Van Rijsingen investigates the 'genomic portraits' by artists Marc Quinn, Andreas Horlitz and Amirthi Perera, and the ways in which they comment on developments in genomic research. The era of genomics has generated a fundamental new concept of human embodiment and identity to which artists have responded with new forms of portraits, showing a deep and critical understanding of what this new concept implies. Van Rijsingen argues that these works reveal that the body can no longer be considered as a container of individual interiority, and that the concept of interiority itself has perhaps become obsolete. DNA 'knowledge' cannot be located somewhere under the surface of the closed skin, at a specific location within the organism, but is produced across membranes constantly in interpretative acts. Moreover, by playfully deconstructing distinctions between scientific and artistic representation, between representation and frame, and between mirror, matter and script, these works make us aware that the body has itself become a medium, the carrier of a genetic text.

CHAPTER ONE

LEONARDO AND FEMALE INTERIORITY

Robert Zwijnenberg

Introduction

Throughout his life Leonardo da Vinci (1452–1519) has engaged in anatomical research. Most likely he also performed several dissections of both human and animal remains. Numerous of his anatomical drawings still exist, the majority of which are characterised by their breathtaking beauty on account of his brilliant control of the art of drawing.

The years between 1504 and 1509 mark a period in which Leonardo was highly devoted to his anatomical studies, as had been true of the years 1487–1495. Even more than in his earlier anatomical drawings he now aspired to put as much knowledge and insight as possible into a single drawing. As such, these later anatomical drawings provide a synthesis of his knowledge resulting from several dissections in combination with traditional notions about the body. His 1507 drawing of the female body (fig. 1.1) constitutes the apotheosis of this synthetic method of representation. It is also the culmination of his visceral studies. To compose this drawing, he consulted drawings of separate organs he made earlier, which he copied unchanged onto this sheet. In doing so, he applied an exceptional variety of drawing techniques. He drew some organs (such as the trachea) in full relief, while others, like the uterus, are represented in transparency; the heart he rendered in cross-section. In other words, to emphasize the specific functions of the organs, which according to Leonardo express themselves in their forms, he drew different perspectives of them: the transparency of the womb shows the cavity in which the child can grow, the way in which the trachea is drawn stresses its pipe-like shape, and the heart's cross-section is eminently suited to demonstrate Leonardo's traditional conception according to which there are two chambers and no atria. He thus emphasized the complexity of the female interior. In its totality, this sheet summarizes Leonardo's conception of the female body. At the same time the many fine details clearly indicate how he wavered between received wisdom and observed facts.

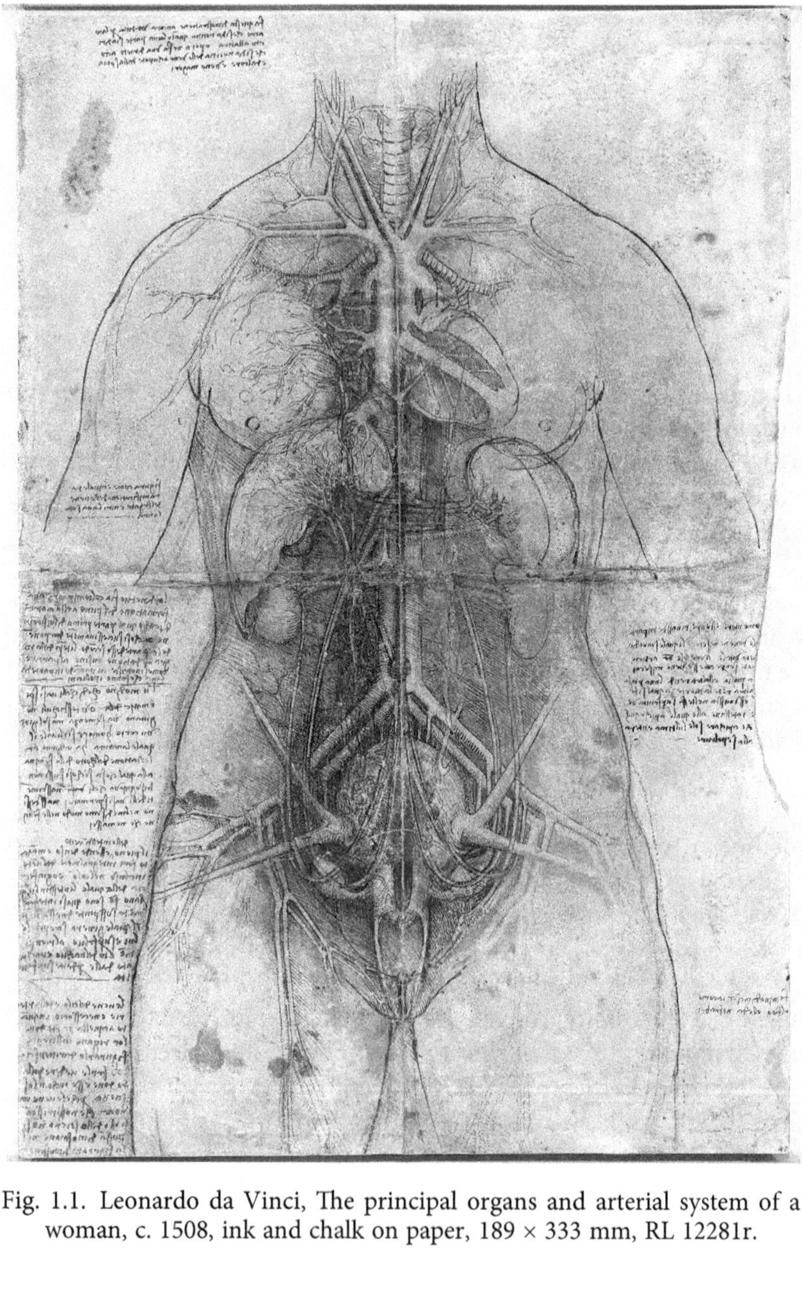

Fig. 1.1. Leonardo da Vinci, The principal organs and arterial system of a woman, c. 1508, ink and chalk on paper, 189 × 333 mm, RL 12281r.

However, the beauty or visual attractiveness of this drawing of this female *situs* figure cannot hide that the drawing is no more than a kind of two-dimensional map of the female body, with three-dimensional depictions of organs that are not very well integrated into this two-dimensional scheme. What is lacking is a convincing depiction of anatomical space: the inner body as an autonomous space with its own specific spatial, physical and emotional qualities.

A few years later Leonardo's anatomical drawing techniques have changed so much that it would no longer be possible for him to make such a *situs* figure. This change, as I will argue, also resulted in a more convincing depiction of anatomical space.

Elsewhere I have shown that Leonardo's new way of anatomical depiction after 1510 is not directly linked with particular anatomical insights, but with notions about the clearest and most profitable way of representing an object—notions which did not originate from anatomical knowledge in the first place. It was in fact the confluence of anatomical knowledge and a new way of drawing that have led to Leonardo's remarkable progress in anatomy after 1510.[1]

What are these new drawing techniques that Leonardo applied after 1510? First, he employed the exploded view, and secondly, he made drawings of one part of the body from different points of view, often combined with the technique of transparency. Leonardo may have learned the use of exploded views from the handbooks by Francesco di Giorgio Martini, an engineer. In his studies of mechanics Leonardo swiftly assimilated the techniques of his predecessors, such as the exploded view and cut-away drawings. Only after 1510 he introduced these drawing techniques in his anatomical drawings.

From his manuscripts it is obvious that all his life Leonardo was struggling with the possibilities and limitations of drawing and painting when it comes to representing nature and the human body. For Leonardo, the main goal of the art of painting is to move the beholder. In his manuscripts, he underlines repeatedly the importance of the relationship between the movements of the soul and the movements of the body, between emotions and gestures, to move the beholder of a painting. We can know only the inner movements of the human body because it has an outer surface. The perceptible body as outer surface

[1] For a more elaborate description, see Zwijnenberg 1999, Chapter 6 "The Anatomical Studies".

of the inner body—that is, of the soul—is indicative of the gesture of a person's life in general. In drawing and painting, the inner movements of the body are only accessible by emotions represented through specific gestures and movements of the members of the body. Or, in other words: only the depiction of gestures and movements makes it possible to represent the outer body as something with an emotional inside. The soul can only be represented in its workings on the body's exterior.

Of course, the interior of the human body comprises more than emotion or the soul. In his *Trattato della Pittura*, time and again Leonardo reiterates the need for the painter to have anatomical knowledge. Only if a painter has extensive knowledge of the anatomical and physiological structures and phenomena of the human body he is capable of representing the human body's exterior faithfully. A convincing depiction of the human body thus necessarily involves a fusion of the emotional and anatomical interior of the body in a representation of the body's exterior.

In his *Trattato della Pittura* Leonardo writes that 'truly, painting does not extend beyond the surface by which the body, the figure of any perceptible thing, is feigned." (TdP, 3, in Farago 1992, 181). At first glance this observation seems a truism. It is evident that a body can only be represented realistically through its surface. But also the interior of a body can only be represented as surface. Each interior element of a body has walls or surfaces; Leonardo's anatomical drawings—but also his mechanical drawings—reveal this much instantly. With his emphasis on the fact that painting is exclusively about surfaces, Leonardo in fact stresses their importance, as well as that painters ought to pay much attention to them. Each represented surface should be an 'emotionally charged' surface, one that expresses both the body's internal structure and its emotional core. Only surfaces that are infused by emotion can realize the goal of the art of painting, which is to move the beholder.

In this essay I will argue that Leonardo tried to pursue his pictorial objective that a surface had to be emotionally charged also in his drawings of the anatomical interior spaces of the human body. The *situs* figure from 1507 reveals that Leonardo had not yet accomplished this goal at that point. However, as suggested above (and as shows from the *situs* figure), closed-off anatomical spaces can only be represented as surfaces. How can the surface of the anatomical space express (inner) emotion? How can the body's interior be represented as an emotionally charged surface?

With the new drawing techniques that Leonardo applied in his anatomical drawings after 1510, he sought to depict, among other things, a strong relationship between surface and emotion. The depiction of the human body's anatomical space he now understood to be a matter of emotions, which is an important element that led to Leonardo's remarkable progress in anatomy after 1510.

A drawing of a foetus

One of his best known drawings in which he did realize this relationship between surface and emotion is a drawing of a foetus in the uterus from about 1511 (fig. 1.2). Leonardo supposedly made this drawing following a real dissection of a pregnant woman. From the viewpoint of our present knowledge however the drawing is not entirely correct: the cotyledons are those of a cow, not of a woman. The drawing has a highly emotional impact because of the way in which Leonardo pictured the full-grown foetus. It is in a stooped position, its hands before the eyes, in a cavity it just fits in, which strengthens the impression of containment and seclusion. Put differently, the foetus is depicted in such a way that its inner being, or its emotional response to its confinement in the enclosed space of the uterus, is expressed in the gestures and movements of its body. In this way, Leonardo can render the inner space of the uterus visually accessible to the beholder, in a cut-away drawing; at an emotional level, however, the drawing presents us with the uterus as an enclosed anatomical space, because of our emphatic connection with the emotional response of the foetus to its confinement.

A major part of Leonardo's anatomical research was devoted to the female reproductive organs. This specific interest was not unusual in his time. In her recent book *Secrets of women*, Katherine Park argues that in early-modern times the anatomical interest in the interior of the human body was especially focused on the female anatomy, the womb appearing as *the* privileged object of dissection in medical images and texts. In Italy's patriarchal urban societies of the late Middle Ages, family membership (including the implied financial and political power) was primarily understood "in terms of blood relationship defined by biological descent through the male line. This emphasis on paternity collided with the realities of conception, gestation and childbirth, all of which foregrounded the mother's contribution to generation and the physical tie between mother and child." (Park 2006, 25) Men were never sure if their child was really theirs. Fatherhood interpreted this way

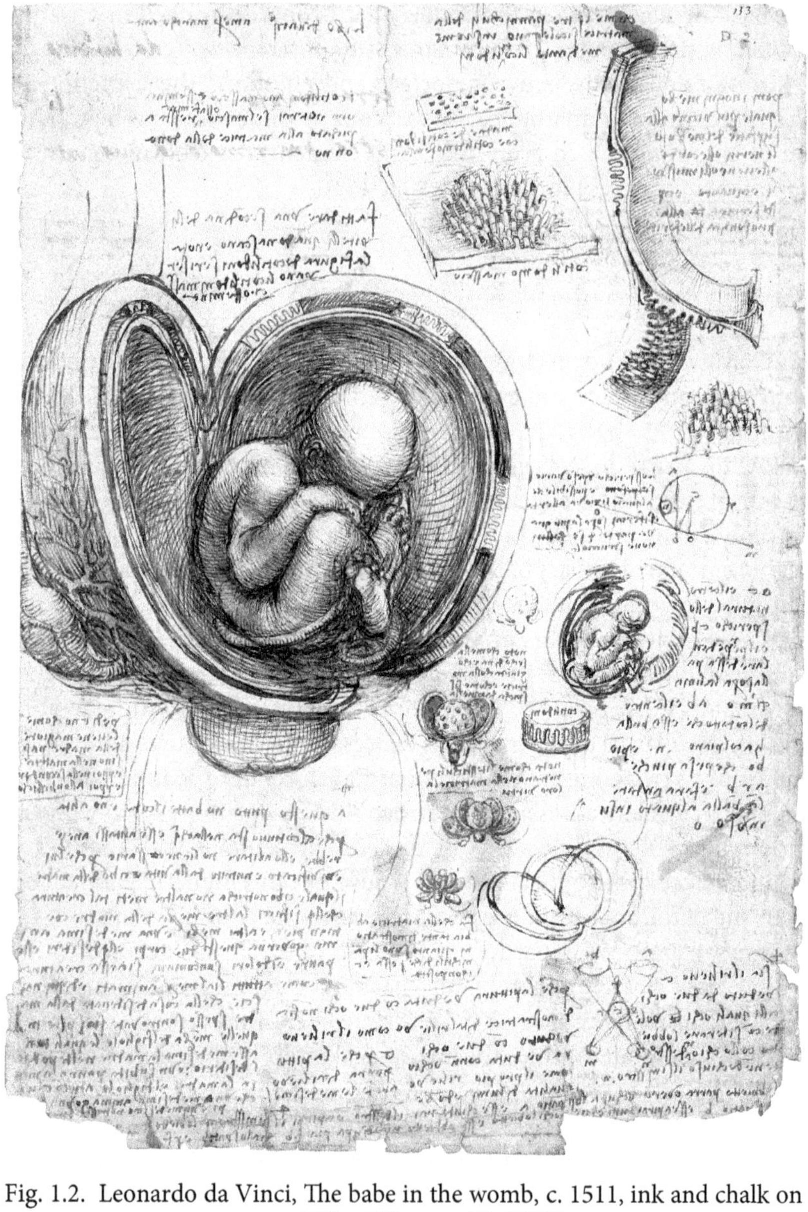

Fig. 1.2. Leonardo da Vinci, The babe in the womb, c. 1511, ink and chalk on paper, 305 × 220 mm, RL 19102r.

fully depends on the sexual fidelity of the woman. There are numerous stories and jokes about the reliability of women in this respect, such as in Boccaccio's *Decamerone.* In his *Il Corbaccio*, also known as *Il laberinto d'amore*, Boccaccio compares the vagina to a labyrinth, and it is made clear that in women, just as in the labyrinth, a great danger is lurking behind a seductive, initially impenetrable façade. In another text, he describes the vagina of his wife as an "obscure valley", and "a great gulf," an "enormous port accommodating many ships at once" and as a "gaping mouth." (Doob 1990, 170)

The uncertainty of fatherhood led anatomists, philosophers, lawyers and theologians alike to pay excessive attention to the uterus—the dark, impenetrable space in which the ties between father and child originate, the sex is determined and the body is formed. Early-modern male authors writing about "the secrets of women" implied that women had access to knowledge concerning sexuality and generation that man did not have, but others would also suggest that these secrets of women are inaccessible to both man and woman. Of course, solid anatomical and physiological knowledge about the interiority of the female body can only be revealed through anatomical dissection.

Leonardo's research into procreation perfectly fits into this context of early-modern notions about fatherhood and the associated misogyny. This is also evidenced by one of his other drawings from about 1508 representing the outer female reproductive organs (fig. 1.3). This drawing is one of Leonardo's most striking studies of a woman's urinogenital system, not in the least because of the confusing combination of distance and directness, underlined by the scratchy way of drawing, which hardly indicates an affectionate approach to the subject. Compared to Leonardo's depictions of women, which are always rife with lyrical beauty and ambiguity (as in *Mona Lisa* or *The virgin and child with Saint Ann*), this drawing is an unexpected, frontal and unavoidable confrontation. We seem to look at the visual pendant of the mysterious smile that is characteristic of Leonardo's portraits of women. The drawing reminds us also of the gaping mouths of the warriors that Leonardo drew in preparation of the now lost Anghiari frescoes.

The drawing of woman's reproductive organ, just like the drawing of the foetus, unmistakably has an emotional load. The fierceness of the drawing discloses an emotion that cannot be easily understood immediately, especially when we compare its style with the usual, more cautious style of his other anatomical drawings. This drawing however may be considered the visual counterpoint of the drawing of

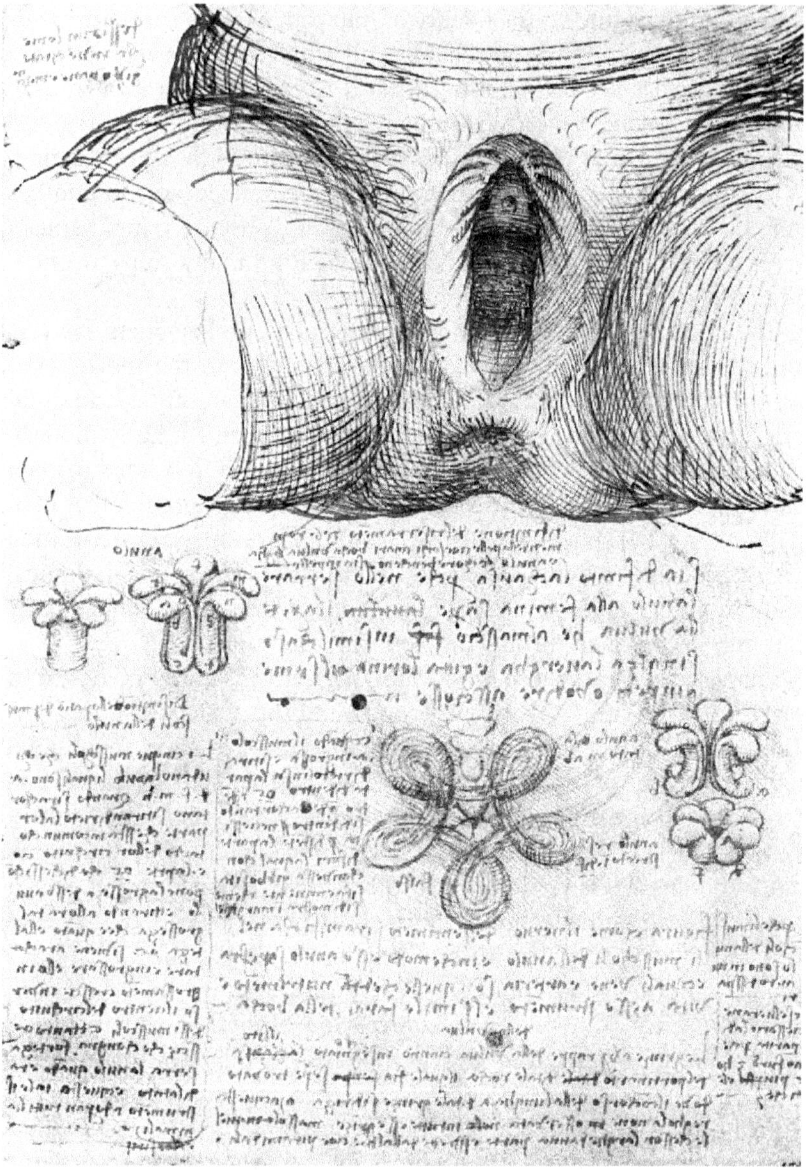

Fig. 1.3. Leonardo da Vinci, The outer genitals of a woman, c. 1508–09, ink and chalk on paper, 191 × 138 mm, RL 19095r.

the foetus because both show us Leonardo reflecting on how to depict the emotional surface and value of the inner female body.

For a proper understanding of Leonardo's possible feelings regarding this drawing, I draw a comparison between this drawing—as well as the one of the foetus—and one of his paintings, *The Virgin of the Rocks* (1483–1486; fig. 1.4), which is now in the Louvre. In so doing I shall not merely explain Leonardo's fascination for the outer and inner female reproductive organs from the anatomical interest that was fed by patriarchal notions, but rather situate this fascination in the much wider perspective of his natural-philosophical way of thinking. Moreover, we also need to assess this drawing from the perspective of his metapictorial thinking, his quest for how, with which methods and drawing techniques, to represent nature and the human body faithfully. Finally I will discuss how this drawing is connected to his thoughts about the relationship between the inner and outer body and emotions. To Leonardo, the depiction of the human body's anatomical space became a matter of emotions, including those about the physical act of dissection.

In this respect it is relevant to note that in Leonardo's view, all things are interconnected: his anatomical research of the female reproductive organs is related to his theological and natural-philosophical views on the mystery of human origins and the origins of natural life in general, which he explored and developed in his manuscripts as well as in his paintings and drawings.

The Virgin of the Rocks

In 1483 the Milanese Brotherhood of Immaculate Conception commissioned Leonardo to produce an altar-piece for the chapel of the Concezione della Vergine in the San Francesco Grande in Milan. This was the start of a series of intrigues, still partly unexplained, involving Leonardo and the Brotherhood. As a result we have two versions of *The Virgin of the Rocks*. The painting, which is now in the Louvre, is assumed to be the one Leonardo started with in 1483. Unsatisfied with the payment he received, Leonardo did not complete the work and he left Milan in 1499 or 1500. Six years later he came back to this city and committed himself to completing the painting, under the additional condition that he would also produce a copy. Most probably this copy or second version—which is now in the National Gallery in

Fig. 1.4. Leonardo da Vinci, *The Virgin of the Rocks*, about 1491–1508, Oil on wood, 189.5 × 120 cm, National Gallery London.

London—came into the hands of the Brotherhood by the end of 1508, whereas the first version got sold to the king of France, the ruler of Milan at the time.

The Brotherhood is likely to have been surprised about Leonardo's work. The contract only mentioned the Virgin Mary and the apostles. St. John was not included. Anyhow, Leonardo's painting is in line with the Mary-devotion of his time, which emphasized her Immaculate Conception. At a first glance Leonardo appears to depict a story from the apocryphal gospel according to St. James, of which numerous popular versions had been circulating since the Middle Ages. At the start of Mary's contractions during the flight from Egypt, Joseph finds a cave, so that Mary can give birth to Jesus. Elisabeth, St. John's mother, is fleeing from Herod's soldiers too, and she reaches the same mountain area. It opens up to her while she is being guarded by an angel. Leonardo's painting represents a meeting between Jesus and St. John in this same mountain region. The flora in the painting symbolises future events: the iris and the palm leaves symbolise Mary and the pool in the foreground refers to baptism. The cave-like surroundings might refer to the Canticles (Songs of Solomon), much favoured by the Brothers of Immaculate Conception for metaphors with regard to Mary: "*My dove in the chasm, hidden in the mountain side, show me your face*" (Canticles 2:14). Such interpretations of this work of art, developed, amongst others, by the art historian Martin Kemp in his book *Leonardo da Vinci* (1981) are really convincing indeed, but they do not come closer to the deeper significance this painting seems to radiate.

In connection with the first version of *The Virgin of the Rocks* I will discuss some texts in Leonardo's manuscripts from 1480 onwards that will shed some new light on this painting, as well as on his anatomical drawings. These texts emphasize the transforming and generative aspects of earth—the continuous process of origin and decay, life and death.[2] In one of the texts Leonardo expresses this as follows: "The hope and the desire of returning home and to one's former state is like the moth to the light, and that the man who with constant longing awaits with joy each new spring, each new summer, each new month and each new year—deeming that the things he longs for are ever too late in

[2] The passages involved are R. 994 (156v), R. 1162 (156v), R. 1217 (156r), R. 1218 (155v), R. 1249 (156v) and R. 1339 (155r). All texts date from ca. 1480, except R. 1219, which is from 1493–4. I quote from the anthology of Leonardo's writings by Richter (1883). References to Richter are indicated by a 'R.' followed by a section number.

coming—does not perceive that he is longing for his own destruction. But this desire is the very quintessence [...] inseparable from nature, and man is the image of the world" (R. 1162). On the same page this text is preceded by a text from a later period (about 1493–4). Leonardo assures that on the one hand nature enjoys creating new life and new species, but that on the other hand it will employ its powers to destroy life again (for instance with poisonous and noxious vapours). He concludes: "This earth therefore seeks to lose its life, desiring only continual reproduction (....) animals are the image of the world" (R. 1219). On the front of this page Leonardo describes a terrible storm at sea. Dolphins and tunas get victimised. After that he pensively reports on his finding a fossil fish in a "cave-like and winding hole": "destroyed by time you are lying here patiently in this secluded place, with your bones stripped of their meat, to serve as a landing point and support for the mountain above you." (R. 1217). But the most beautiful and penetrating description is on another page:

> After some wandering between bleak rocks I came to the entrance of a cave. I stopped for some time, surprised and unfamiliar with something like this. I arched my back and rested my left hand on my knee. I held my right hand against the light, over my down-cast and contracted eye brows. I moved a few times to and fro in order to see if I could see something inside the cave, but the great darkness that prevailed over there was an obstacle. After standing there for a while suddenly two things emerged in me, fear and desire—fear for the dark and threatening cave, desire to see whether some wonderful thing was there. (R. 1339)

This text is very special indeed because, exceptionally, it reveals something to us of Leonardo's personal commitment—his emotions. His texts that I quoted also seem to suggest that *The Virgin of the rocks* has its roots in a tradition that is much older than Christianity, viz. the notion of earth as the universal mother. This suggestion also appeared in Walter Pater's essay on Leonardo da Vinci, part of his famous work *The Renaissance* (1873). Pater observed that in many of Leonardo's paintings the women are represented with water in the background, both versions of *The Annunciation, Ginevra de'Benci, Mona Lisa, The Madonna of the Yarnwinder, The Virgin and Child with Saint Ann, Leda and the Swan* and both versions of *The Virgin of the Rocks*. Pater mentions the "solemn effects of moving water" and he describes the women as clairvoyants who not belong to the Christian family. Furthermore, his description of *Mona Lisa* includes a reference to the sibyl: "She is older than the rocks among which she sits, like a vampire of the grave; and has been a diver in deep seas. . . ." (Pater 1980, 91)

In her article "On looking into the abyss: Leonardo's Virgin on the rocks" (1997), Regina Stefaniak advocates an interpretation of this painting that seems to elaborate Pater's views. She connects the water in the background of the painting with the "primeval flood" in Genesis 1:2, and the whole picture with Proverbs 8:22–25:

> *The Lord has acquired me before all other things; when He started his creation He first created me. I was made in the beginning, even before everything, even before the earth was shaped. When there were no oceans I was originated, even before the springs with their flows of water. Before the mountains were erected I was originated, even before there were hills.*

The expression "primeval flood" referred to in both Bible texts is a translation from the Greek abyss, which literally means "bottomless". In commentaries from the 7th century on the Latin Vulgate translation the notion is extended to *matrix abyssus*. In Late Latin matrix means uterus, source, origin, or cause. From this matrix all rivers and springs on earth originate and they also flow back into it. Although it is not possible here to account for Stefeniak's interpretation in great detail, she demonstrates convincingly that Leonardo evokes the primeval flood in this painting, offering the theological interpretation of the Virgin Mary's uterus as the *matrix abyssus*. In the painting we visually enter this matrix. We are looking into Virgin Mary's uterus, in which Christ and St. John are united. As these holy persons are pictured too large in proportion to their surroundings, the same feeling of containment and seclusion comes to mind as in the drawing of the foetus in the uterus. Again, the theological background and complexity of this interpretation require a careful reading of Stefeniak's article in full, but her striking and convincing conclusions essentially boil down to the explanation briefly outlined here.

Fear, desire and anatomy

As I observed above, Leonardo's text on the cave is very special because it reveals us more about his personal commitment, while generally his manuscripts do not show any personal commitment or emotion. This text is special, too, because Leonardo immediately links emotions such as fear and desire to a space that is hidden behind a seductive, initially impenetrable façade, to recall Boccaccio's description of woman in *Il Corbaccio*, a text that Leonardo probably knew.

Moreover, Leonardo's text aptly describes the experience of an anatomist who is about to dissect a body. Anatomy involves a paradoxical situation: the anatomist wants to show the truth of the body, but in order to do so he or she must destroy the object of his or her deep attention. It must be broken into fragments that can never be joined together again into an ordered whole. This rejoining can happen only in a representation, which in fact frequently presents an order which never existed in the real world. We see this, for instance, in Leonardo's drawing of the female body, which I discussed at the beginning of my essay. The main reason for dissecting a body is to expose its structure. But dissection destroys the body, which results in a paradox: anatomy aims at the revelation of order, but it also causes the destruction of order.

We recognize in Leonardo's text the fear and desire of the anatomist to enter the human body, to make the first cut to open up the body to reveal a space that is covered by an emotionally charged surface. In general, in early modern texts the act of cutting open human bodies is described as a daring, frightening and unwarranted deed. In fact, still today medical students consider the first incision in a corpse a barrier to be overcome only through courage.

Turning back to Leonardo's drawing of the outer female reproductive organs, we recognise in the background of the drawing the same contradictory emotions Leonardo experienced when he was in front of the cave: fear and desire. The notion of an interior and an exterior, of a boundary, a threshold that must be crossed—a notion characteristic of the thinking about caves—is emphasized by a note next to the drawing: "the wrinkles or ridges in the folds of the vulva reveal the place of the gate-keepers to us." Leonardo's confronting drawing of the female reproductive organs may refer to the misogynous metaphors of his time. In a different note he also expressed his intense revulsion of sexual intercourse, while in the following phrase one can also recognise the paradoxically misogynous fear of the secrets of woman: "The coital act and the parts that play a role in it are so disgusting that without the beauty of the faces and the magnificence of the participants nature would be without the human species" (R. 1145). If, however, we were to have a second look at the drawings of the foetus and the female sexual organs while taking into account Leonardo's thoughts about the origin and decay in nature, we would conclude that the drawings represent as much a physical and intellectual exercise of the mind as does *The Virgin of the Rocks*. This painting is Leonardo's intellectual and emotional reflection on the theological meaning of the uterus as *matrix abyssus*. In addition, the painting's beholder is invited to come nearer by the Archangel Gabriel's outward gaze and the gesture of his hand. But access to the painting is almost physically obstructed by the pool in the foreground, while the emotional intimacy of the scene also seems to shut out the beholder. As beholder we find ourselves in a similar position as Leonardo in his story about being in front of the cave.

Leonardo's anatomical investigations into the female body and the origin of life in the uterus bring him, in a physical way, into contact with mystical notions about life's origin in the *matrix abyssus*. When drawing, Leonardo is in a literal and figurative sense confronted with the abyss life originates from and returns to. It is as if in the drawing of the outer genitals Leonardo is meditating upon the anxiety, horrors and expectations of anatomy, the fear and desire of entering the human body. These forceful emotions are reflected in the drawing. This means that this drawing embodies the physical aspect of his anatomical study of the female body and the origin of life in the uterus, or the actual act of dissection. Yet the theological and natural philosophical ramifications of the physical encounter with the female interior are of course also at work in this drawing. On account of Leonardo's confusing and thrilling physical and intellectual experience when dissecting a female body, these two drawings are permeated with an emotional impact and meaning that far exceed the limited patriarchal views of his time.

Leonardo belonged to the first anatomists who opened up the body again in pursuit of anatomical knowledge, and in my view he may have been the first anatomist who confronted the problems of depicting the new anatomical experience and knowledge that resulted from these dissections. The late medieval methods of anatomical representation, as in the *situs* figures of Vigevano and Mondino, were clearly inadequate to depict the new anatomical knowledge. The remarkable progress in Leonardo's anatomical research after 1510 that is especially visible in his late anatomical drawings must be related to the fact that he found a way to make the emotions of dissection, of entering a human body, an integral part of the depiction of the inner spaces of the female and of course the male body. He found a way to overcome the pictorial problem that the closed-off space of the human body's interior can only be depicted as a surface, while in his view this surface, to be convincing and true to nature, had to be a surface that was charged with emotion. What he had already achieved in his painting, he now achieved in his anatomical drawings: the body's inner spaces can be represented as closed inner spaces through emotionally charged surfaces.

CHAPTER TWO

ANIMALS INSIDE: ANATOMY, INTERIORITY AND VIRTUE IN THE EARLY MODERN DUTCH REPUBLIC

Rina Knoeff

Sometime early in the eighteenth century, 78-year-old Grietje Willems, a fishmonger's wife and living at the Wester-Schelling, sent for *Dominee* Adriaan Byman and Bailiff Adriaan Klein.[1] Something extraordinary had happened to her, and she was extremely worried. Having been violently sick, anxious and short of breath for some time, she had vomited up a small four-legged creature, which resembled a little dog. It had a head, a mouth, a tongue, legs, a tail and something like an umbilical cord. It was enveloped in a small pocket, resembling a womb. The lady's granddaughter, Jacomijntje Alberts, was with her at the time. As Grietje Willems was feeling very sick, Jacomijntje fetched her mother and several neighbours.

In this period a strange vomiting was seen as a possible indication of sorcery or fraud, so it was a logical step to call in the pastor and the bailiff.[2] They threw the creature in the fire to test it, but it did not burn. Although one might expect this to be a sign of possible witchcraft, nothing happened. Grietje Willems lived on and stood by her story until she died. The dog-like creature was preserved in brandy and olive oil and presented to anatomist Frederik Ruysch (1638–1731) to be examined and exhibited in his anatomical cabinet (Ruysch 1744, vol. 2, 793–796). The cabinet itself does not exist anymore, but we can still see the little "dog" on a picture accompanying the written accounts of Ruysch's collections (fig. 2.1).

[1] I use the term *dominee* as this was the proper title of a pastor of the Dutch Reformed Church.

[2] The story corroborates Claudia Swan's argument that although the early modern 'witch-craze' was not so pronounced in Holland (compared with other European countries), this does not reflect lack of interest in witchcraft (Swan 2005). A similar argument was put forward by Price (Price 1998, 148–149). For witchcraft in the Netherlands, see also Gijswijt-Hofstra and Frijhoff, (eds) (1987).

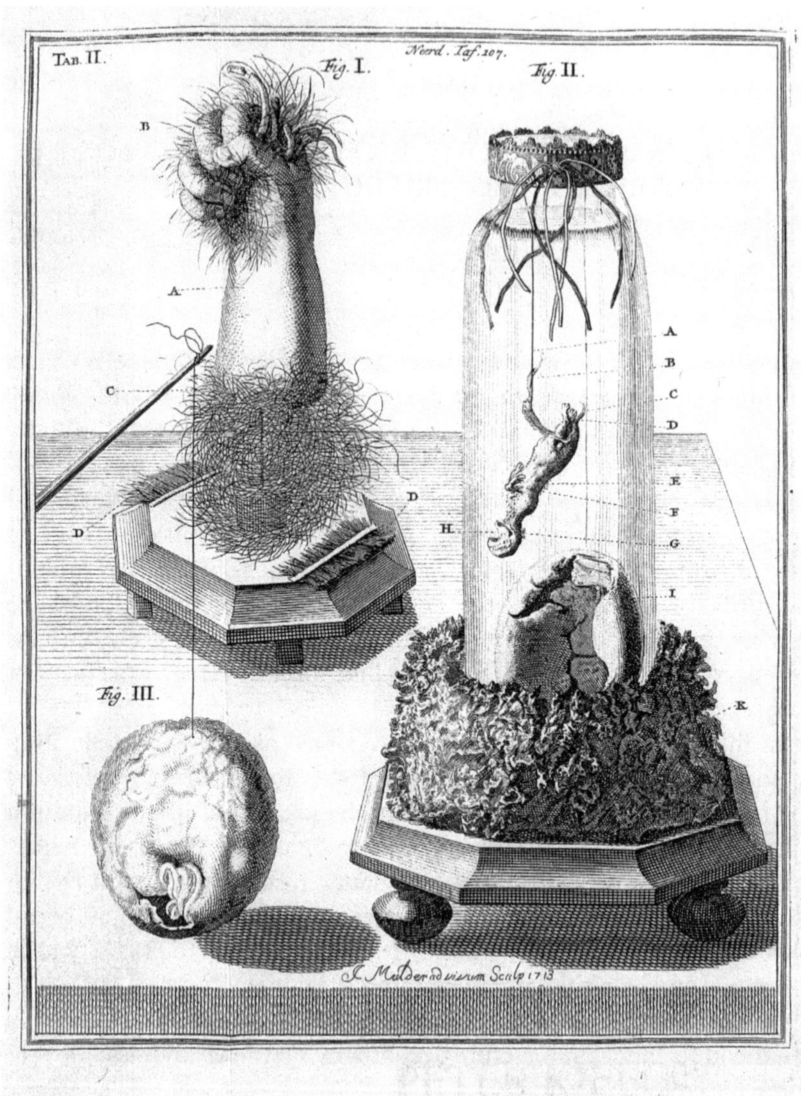

Fig. 2.1. From: Ruysch's *Werken*, volume 2, illustration 107, figures II and III: depiction of the little dog and the back of the 'pocket resembling a womb' in which the animal was found. The engraving is by Josef Mulder, 1713.

What are we to make of Grietje Willems' strange sickness? It is hard to imagine that she really vomited up a little dog, so what status should we ascribe to her case? I shall argue that Grietje Willems' story is but one example—albeit a very striking one—of how the early-moderns tangibly imagined disease and corruption of the inner body in the form of animals living inside them. Nowadays most of us have a clear idea about what our bodies look like inside. This idea is *pictorial* and based on modern visualisation technologies. By comparison, in Grietje Willems' time it was normally impossible to see the inside of a living person. Medics deduced the interior functioning of bodies from the anatomy of corpses. More generally, ideas about the working of the inner body were based on people's imagination and intimate *experience* of their own bodies. Pain, for instance, or noises from within, led to speculations about what could possibly be going on inside. This is not to say that physiological processes were thought about largely in abstract terms. Instead, perhaps even more than today, the early modern image of interior processes was visual, and in the case of Grietje Willems even tangible.

It has been argued that from the sixteenth century onwards, as a result of the "anatomical renaissance", people increasingly imagined their personhood (that is his being mortal, fragile, prone to illness and the product of a divine plan) in anatomical terms (Park 2006, 264).[3] Following this argument, I suggest that it was not only anatomical views, but also stories like the one of Grietje Willems which determined how people experienced their identity. As such these stories are not only interesting in their own right, but also, and perhaps more importantly, represent cultural ideas and anxieties about illness, pain, gender and sexuality. Equally, the stories had a remarkable influence on anatomical research. After all, the animals which surfaced from the body often ended up in anatomical collections, thus visibly revealing the possibly disgusting secrets of the human interior.

In this paper I shall first discuss the early modern phenomenon of animals living and breeding inside human bodies, which was widely known at the time as the "Bosom Serpent". Secondly, I shall look at contemporary physiological explanations of how animals can end up inside humans. In the third and final part of the paper, I shall turn to

[3] See also Sawday (1995). The term 'anatomical renaissance' is from Andrew Cunningham (Cunningham 1997).

the alleged appearance of "animals inside" in the cultural and political context of the early eighteenth-century Dutch Republic.

My focus will be mainly on the work of Dutch anatomist Frederik Ruysch, whose anatomical cabinets were internationally known and admired. By means of a "secret" technique of injecting a wax-like fluid into blood—and other vessels of the body, Ruysch preserved rarities, body parts, the bodies of unborn foetuses and tiny infants, in such a way that they looked alive.[4] They were exhibited and freely accessible to the public. Ruysch's written accounts describing the collections not only reflect what was on show, but also, and particularly in the case of unusual exhibits, tell how Ruysch acquired particular specimens and include medical explanations. And Ruysch is almost nowhere as elaborate as when he discusses the occurrence of animals living and breeding inside the human body.

The Bosom Serpent

Grietje Willems' story is not the only one of its kind. At the time, stories about animals breeding inside human bodies occurred all over Europe. Well versed in ancient legends and miracle stories, people earnestly believed in the possible existence of snakes, frogs, lizards and other animals inhabiting the intestines.[5] Even Hippocrates, in his *De morbis vulgaribus* described how a youth died when a large snake slithered out of his mouth after the boy had drunk a large quantity of strong wine. In the Middle Ages, the phenomenon known as the "Bosom Serpent" was mentioned frequently in legends, medical works and the saints' lives, and even until the twentieth century accounts continued to surface of people suffering from live animals in the stomach. What is more, these accounts were taken very seriously. Although in many instances they turned out to be fraudulent, they remained the subject of earnest medical speculation until well into the eighteenth century.

---

[4] For Ruysch see Luuc Kooijman's excellent biography (Kooijmans 2004) as well as the work of Julie Hansen (Hansen 1996a; Hansen 1996b).

[5] Jan Bondeson collected various stories from all over Europe. (Bondeson 1997, 116–119). For miracle accounts of snakes and frogs living and breeding inside human beings see also Wilhelm (1983). For a more 'medical' discussion of the occurrence of worms inside people see Zedler (1732–1754, vol. 60 (1749), col. 16–21, "Wurm").

Probably the most famous case was that of Catharina Geisslerin, "the toad-vomiting woman of Altenburg". She first vomited a toad in 1642, and this was followed by many more such instances. She claimed that she could feel toads, frogs and lizards running around in her intestines, particularly after drinking milk, which the amphibians liked. After two years of her suffering, one of the animals was sent to anatomist Thomas Bartholin (1616–1680) in Copenhagen, who discovered over thirty black winged insects in the toad's stomach. This seems highly unlikely food for a toad born and bred in the human body, but Bartholin did not conclude that the woman purposely swallowed the toads before vomiting them up. Instead he published an article in which he confirmed the possibility of toads and lizards living inside the stomach (Bartholin 1637). After Catharina Geisslerin's death in 1662, doctors were eager to perform an autopsy, expecting to find all sorts of amphibians within her body. Yet, they did not find any animals of the kind and they ascribed her death to dropsy and inflammation of the liver.[6]

In addition to the case of Grietje Willems, Ruysch encountered a case of strange vomiting on one further occasion. The father of a fourteen-year-old girl called him to attend to his daughter who, having been very ill for years, vomited a kind of slug with two eyes and an umbilical cord (fig. 2.2). The animal was alive and well, because for a long time it kept moving about. Like Bartholin in the case of Catharina Geisslerin, Ruysch did not believe that this was a fraud. He explicitly denied the possibility that the girl had first swallowed the slug and then vomited it. He maintained that this could not possibly have been the case because the animal did not creep like ordinary slugs, but rather moved up and down, thereby proving that it was a *special* kind of slug. Additionally, he argued that the girl had been ill for too long for the animal to have survived in the stomach (Ruysch 1744, vol. 3, 955).

In line with some recent research on the experience of the inner body, I argue that the stories, which were matter of great curiosity, represent early modern ideas and anxieties about human physiology (Porter 1985; Stolberg 2000; Stolberg 2003). Barbara Duden has argued that in the eighteenth century the inside of the living body "could be grasped only as the place of an experienced but invisible flowing" and that therefore an evacuation enabled the doctor to determine what was going on inside (Duden 1991). Moreover, in a recent article on the intimate experience

[6] The full story can be found in Bondeson (1997, 33–34).

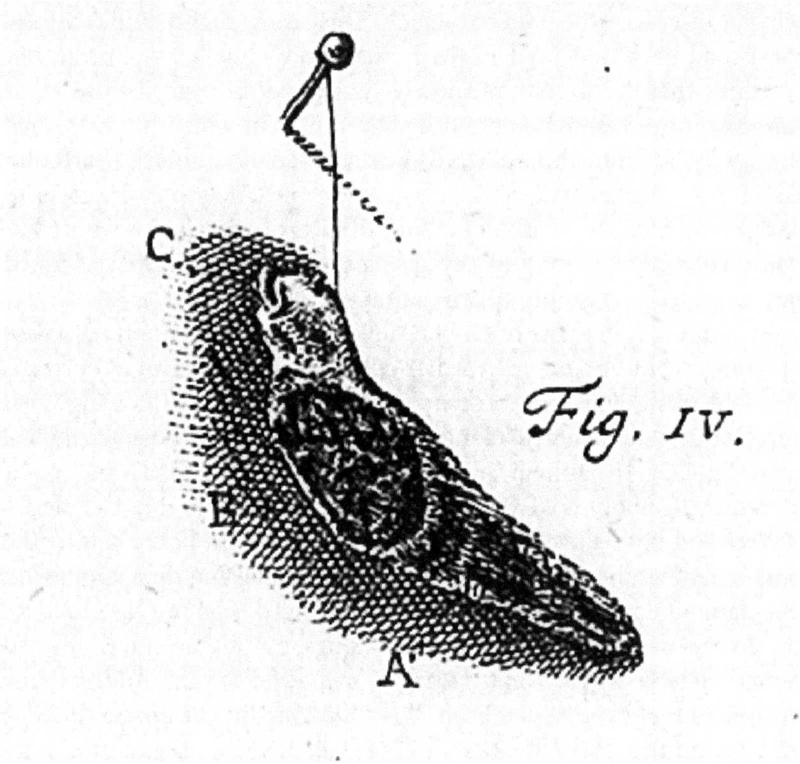

Fig. 2.2. From Ruysch's *Werken*, volume 3, illustration 121, figure IV: Representation of the slug thrown up by the fourteen-year-old girl. The engraving is by Jan Wandelaar.

of the body in the eighteenth century, Severine Pilloud and Micheline Louis-Courvoisier have stated likewise that secretions were "useful for achieving a better understanding of the interior of the body" (Pilloud, Louis Courvoisier 2003, 468). Indeed, historians have so far discussed the hidden nature of urine, blood, stools and vomit, but the rather more extraordinary phenomenon of animals representing the body's inside has rarely been addressed.

While historians like Jan Bondeson in his *A Cabinet of Medical Curiosities* primarily aim at "unveiling the [fraudulent] bosom serpent", I maintain that the stories should be taken seriously. They are examples of what Barbara Stafford calls the early modern "activity of visibilising, or incarnating, the invisible", which "became endowed with a special urgency in early modern art and experimentalisation" (Stafford 1991, 17). My argument, however, is more radical than Stafford's. She has focused narrowly on the Enlightenment invention of *metaphors* in art and medicine, which represent the unknown depths of the body. In her view, stories about animals afflicting the human body should be understood figuratively. Satirical images showing animals attacking the body were produced in the mind, she argues, they were "irreverently compounded from new or complex ideas not found in empirical reality [but] nonetheless [corresponding] to the contradictory 'feel' of that reality" (Stafford 1991, 192). Similarly, Robert Jütte has understood stories about animals living and breeding inside the body in the context of popular belief and folklore, pointing to the early modern preoccupation with "imagination" as a possible explanation for the alleged monstrous growth of animals inside humans (Jütte 1996, 193–215).[7]

I would maintain, however, that stories about, and pictures of, animals inside humans were not just mental images summoned to help people deal with pain and discomfort. They were based on a *real* fear of animals breeding inside people: to the early moderns grotesque representations of the interior body, such as Grietje Willems' vomited dog, were visible and tangible objects rather than mere imaginings of the mind. Even though not many people actually witnessed the vomiting episode, *anyone* could visit Ruysch's anatomical cabinets and *see* the strange creature. They could almost touch it and, as it were, relive the horror of the sickness of Grietje Willems. Animals surfacing from

[7] For popular images of disease and pain as animals living and breeding inside see also Bargheer (1931, 416–426).

the inside were not dismissed as the fancies of the uneducated, but keenly investigated by anatomists and physiologists. In other words, the animals were taken into the medical domain, and in most cases it was explicitly denied that the stories were fraudulent or related to supernatural powers (note, for example, the testimonies of the pastor and the bailiff in the case of Grietje Willems). As products of the interior body, the animals tangibly represented and made visible what was going on inside. However, before discussing how this came about, I shall turn first to contemporary medical explanations of the phenomenon.

Medical explanations

Several theories about how animals could live and breed inside the human body circulated in the eighteenth century.[8] The oldest explanations were closely related to legend and included the possibility of snakes and lizards creeping down the throats of sleeping people. Another explanation was that sorcerers and witches deliberately put reptiles inside their victims, which perhaps explains the value attached to the testimony of a man of the church in the case of Grietje Willems.

Other, more medical, explanations included the originally Aristotelian theory of spontaneous generation, which posits the formation of worms and other animals in the corrupted humours of the intestines. Drinking of polluted water was also considered a cause for animals growing inside. Until the eighteenth century, the latter explanation was accepted by many natural philosophers and physicians. Carl Linnaeus, for instance, assumed that by drinking polluted water, people could accidentally swallow spawn which might attach itself to the stomach and develop into amphibians, which could live and breed within the intestinal tracts (Bondeon 1997, 30, 36, 38; Linnaeus 1995 (1732)).

The invention of the microscope, and the ensuing discovery of all kinds of tiny animals living in the waters of ponds and infusions reinforced the view that animals might well end up inside human beings. Indeed, when in the 1660s the microscope was applied in medical

[8] No doubt these stories were based on 'true' observations of worms in human (particularly children's) excrement. Herman Boerhaave, in his influential medical textbooks, lists worms among children's diseases and states that they surface in excrement, vomit and even mucus. He maintains that worms seldom appear in adults. (Boerhaave 1741, aphorisms 1359–1373).

research, small creatures were discovered living in various organs and secretions. Ruysch, for instance, discovered worms in the arteries and liver of horses, in the gall bladder of sheep, and in the kidneys of dogs and man. He had no doubt, moreover, that worms can also be found in a man's brain (Ruysch 1744, vol. 1, 105). For Ruysch, as well as for other medical men, the microscope provided convincing evidence for the existence of independent life inside the body.[9]

Even though it was virtually impossible to distinguish many details within these creatures clearly (even with a microscope), they were naturally assumed to possess animal characteristics. The anatomist Govert Bidloo, for instance, argued in a discussion with natural philosopher Antonie van Leeuwenhoek, that the liver fluke, a small worm living in the liver of sheep, *must* have eyes simply because all animals (as it were by definition) have them (Bidloo 1698). In this sense it is highly questionable whether Ruysch preserved a little dog or some creature to which he ascribed the characteristics of a dog. In any case, it is understandable that microscopic research fuelled speculations about animals living inside the body and increased apprehension about what might possibly be going on inside; in fact, they raised the likelihood that little animals could be found in vomit.

Unlike Linnaeus, Ruysch considered it improbable that worms entered the body by the mouth. He maintained that fermentation processes in the stomach transformed everything and killed all alien life. He also denied the possibility of worms invading the body through pores in the skin or through the respiratory organs. Moreover, he reasoned that the animals had never been observed outside the human body, which left the conclusion that they must have been generated by the body itself (Ruysch 1744, vol. 1, 105). Ruysch's view fits eighteenth-century speculations about generation, and the possibility of independent life inside the human body. Although many natural philosophers and physicians no longer believed in the theory of spontaneous generation, other theories were brought forward that still included the possibility of animals breeding inside the body. The popular theory of preformation, for instance, held that "all the living things there were had to be, had in fact been created by God at Creation, and that, encapsulated within the first parent all future generations were present" (Gasking

[9] For the use of the microscope in early modern anatomy, see Fournier (1991); Wilson (1995); Ruestow (1996).

1967, 42). In accordance with this idea the theory was brought forward that the body carries within itself "a series of animals becoming smaller and smaller to infinity", unfolding like a Russian babushka, as it were (Malebranche in Gasking 1967, 46). Yet the preformation theory only allowed for animal bodies to generate bodies like themselves, so the puzzling question remained of how animals could possibly end up inside humans.

Other mid-eighteenth-century theories solved the problem. The well-known and influential French natural philosopher Georges-Louis Leclerc, Comte de Buffon, for instance, argued that there was a sharp divide between organic and inorganic matter. The former contained vital principles that, having been taken into a body, could reproduce and cause the growth of other organisms. These did not necessarily have to be the same as the parenting body. Indeed, Buffon suggested that worms and other parasitic animals in the stomach and intestines might arise from surplus vital particles that had not been absorbed into the body but had instead been transformed into other animate beings. Albrecht von Haller assumed the presence of spontaneously produced *animalculae* and concluded that the foetus was present in the female egg and had to be stimulated into life by the male semen. From von Haller's theory it was but a small step to hold that the presence of an animal egg in a human body could cause the growth of an animal inside.

Ruysch, following a French essay on the matter, also believed that all animal and human life originated in eggs.[10] These eggs, he argued, were present in the male sperm. He wrote that "there is no animal that is not born from the egg of a similar animal, and that in the male sperm of man are hidden many eggs of other animals, which, after some time, finally hatch out and grow. Often living animals surface" (Ruysch 1744, vol. 3, 1071). Thus, in the case of the fourteen-year-old girl, Ruysch suggested that the slug she threw up originated from eggs present in her father's sperm. He argued that these eggs had remained in her body after she was conceived, and had hatched out in due course. An animal of this kind could grow inside and cause a disease.

Ruysch considered the occurrence of animals in the body in much the same way as a tumour or a monstrous birth, which was exactly how he explained the presence of a little dog in Grietje Willems' stomach

[10] Ruysch referred to (Andry 1700). As a result of the book Andry was nicknamed 'Man of Worms' or 'Wormy'.

(Ruysch, 1744, vol. 2, 956). He suggested that perhaps the pocket which contained the little dog was what was known as a 'pulptumour' (*papgezwel* or *Atheroma*)—an unnatural growth often containing teeth, hair or bones.[11] Ruysch, however, did not consider the matter any further and assigned the phenomenon to the domain of the philosophers, "the researchers of secrets".[12] He wrote:

> When anyone should ask how these animals invade these hollow spaces, I would ask in return how a bundle of worms can end up in the artery below the heart? I have seen and described it. How can one explain the row of teeth in the 'eggnest' [ovaries] of a woman? Or the presence of worms of thirty *el* [= 20,7 meter] in man's body? Or what about the shin bone of a child in a 'pulptumour' [*papgezwel*]? Or the presence of thirty fat caterpillars in a sheep's brain? And where did the vomited little animal in the small pocket come from? All these things I have nevertheless seen and kept in my cabinets (...) The whole of natural history cannot explain the way these miracles are possible (Ruysch 1744, vol. 3, 1070).

Picturing the Inside Body

Central to all stories about animals living inside the body was a general anxiety about animals creeping inside and thus corrupting existing structures. This is not surprising since the devastating effects of pile worms were visible everywhere in early eighteenth-century Dutch society. The animals ate their way through the dikes, causing them

[11] Ruysch exhibited several such puzzling tumours in his anatomical cabinets.

[12] This was not unusual. At the time, philosophy—and particularly natural philosophy—was concerned with the 'nature of things' and therefore closely related to physiology (the term derived from the Greek *physis*). Andrew Cunningham has argued that early modern physiology is based on the knowledge acquired from dissection and anatomical research and is concerned with the causes of phenomena. In other words, while anatomy was considered a manual discipline (art), physiology was seen as a theoretical discipline (science) and indispensable to medicine. 'Only by contemplation, by reasoning, will we understand the true causes of the nature of the human. For it is only by thought that we can understand the nature of the sub-visible constituents of the body and how the parts co-operate in bringing about the great functions such as digestion and generation. Physiology is the rational—not empirical, nor experimental—discipline which lies between anatomy and medicine'. Ruysch was originally trained as an apothecary and was never much interested in physiological questions. For him it was enough to preserve and exhibit the 'animals' and he was more than happy to leave the question of what could have caused the illness to 'the philosophers' (Cunningham 2002). For Ruysch's anatomical descriptions of the body as opposed to a more "theoretical" approach see Knoeff (2006).

to collapse which led to severe floods in 1731. Moreover, they were gnawing away the wooden pillars upon which Dutch houses were built, thereby endangering the existence of thriving Dutch cities like Amsterdam. Many took the worms as a sign of divine punishment for sexual deviancy and sin and pamphlets appeared with titles such as *The Worm as a Warning to the Feckless and Sinful Netherlands* and *The Finger of God, Or Holland and Zeeland in Great Need from this Hitherto Unheard Plague of Worms.* The authors argued that the worms had been "custom-made by the Almighty for the express purpose of punishing a stiff-necked people steeped in filth and sin" (Schama 1991, 607).[13] After all, biblical stories often referred to God sending animals to punish His people, so why should He not do so in the early modern Dutch Republic? People reacted feverishly, and before natural historians could dare to explain the phenomenon in "biological" terms, "special fast days were announced (one wit adding his hope that the worm might join the fast) and acts of public collective contrition like special donations for the poor" (Schama 1991, 607).

Worms, in other words, were generally associated with moral corruption and the consequent ruin of the material culture of Dutch society and its protections against the sea. The image of worms undermining the Republic was all the more poignant because ever since 1588, when the United Provinces had become a republic, the Dutch had continuously struggled to preserve the delicate balance of liberty, stability, virtue, and prosperity, and to guard the Republic against political decay. Radical intellectual movements such as Socinianism and Cartesianism were perceived as a threat to political stability (Israel 1995, 791). Pile worms were a further and perhaps even greater danger. They embodied the corruption of the Republic and the consequent disappearance of land and goods into the sea.

As the devastating effects of the pile worms were clear for all to see, stories about animals living and breeding inside humans made people urgently aware that animals could even corrupt people themselves. As such, accounts like that of Grietje Willems had a particular function in eighteenth-century Dutch society. Just as worms materially symbolised the collapsing structures of "diseased" dikes and houses, the little dog literally represented the pain and discomfort of the diseased body. Moreover, the contemporary association of pile worms with immoral-

[13] For the disastrous effects of worms see also van der Heijden (2004).

ity and sexual sins gave the occurrence of animals in female bodies a different quality to similar (albeit much less frequent) manifestations in male bodies. I shall highlight these two aspects ((1) worms embodying pain and discomfort and (2) the occurrence of worms in relation to the female body) in the last section of this paper.

Pain and Discomfort

One significance of the "Bosom Serpent" relates to particular experiences of pain and distress. Stolberg has argued that the early moderns often experienced pain as something invading the body from the outside Stolberg 2003, 43). The concept of "animals inside" is a very clear illustration of this. The animals visibly and tangibly represented interior corruption, after all, the emerging animals explained the extreme discomfort that Grietje Willems and the fourteen-year old girl felt before vomiting. Illness and pain were reduced to the manageable proportions of a small creature living inside. The fact that both Grietje Willems and the fourteen-year-old girl felt better after throwing up reinforces my suggestion.

Ruysch's discovery of worms and flies inside human bones further explains the transformation of the early modern experience of illness into the visible and palpable phenomenon of animals breeding inside. Although the skeleton had always been a topic of anatomical research, the structure of the bones themselves had been studied much less. Ruysch, after some discussions with his friend Herman Boerhaave, decided to remedy the situation. He relates that the "very bright artist" (*konstenaar*) whom he hired to cut the bones lengthwise discovered and showed him the remains [*chrysalides, vel nymphae*] of worms and flies within the bones (fig. 2.3). The pupae were so small and hidden away that they almost escaped Ruysch's attention, but, Ruysch writes, his extensive knowledge of the natural history of bloodless animals allowed him to recognise the remains for what they were. He immediately started to investigate the bones of arms, thighs and shoulders of both adults and children, finding not only pupae, but also the worms and flies themselves. And as the bones were intact on all sides (Ruysch declared that he did not find even the tiniest of holes), Ruysch considered it impossible for the creatures to have entered the bones after death. Furthermore, he argued to sceptics that he discovered the animals in the arm of a one-year-old child, which implied that they

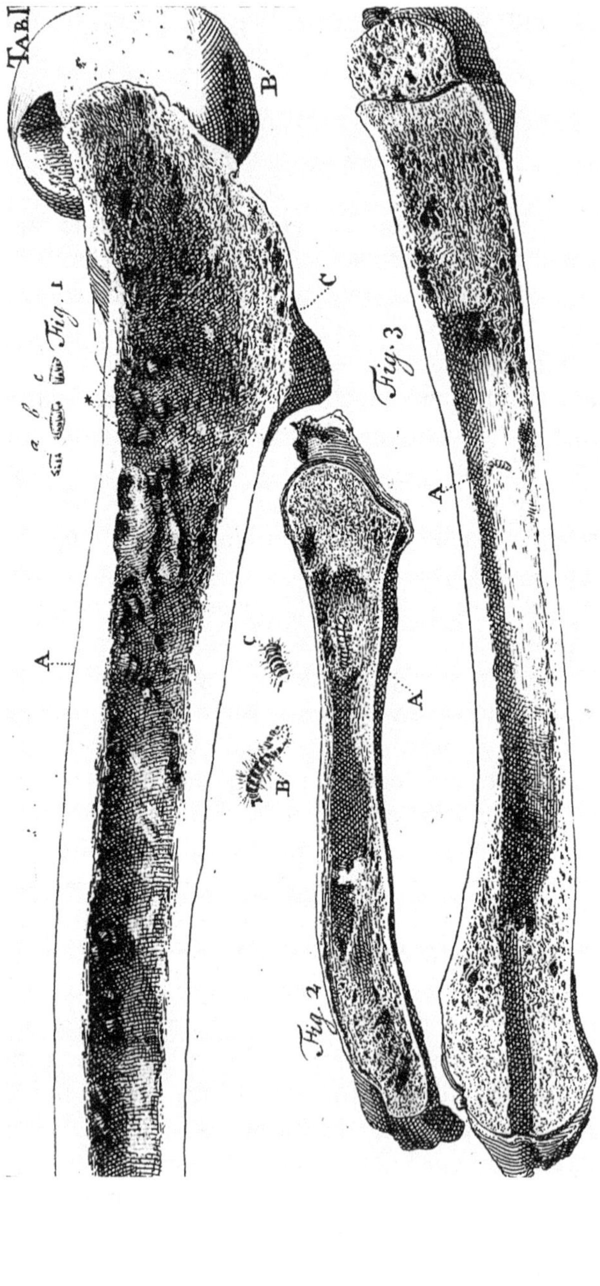

Fig. 2.3. From Ruysch's *Werken*, volume 3, illustration 125, figure I–III: Representations of worms found inside human bones. The engravings are by Jan Wandelaar.

must have been present from the time of birth (Ruysch 1744, vol. 3, 1069–1072, 1086–1088).

More than in the case of Grietje Willems, Ruysch explicitly linked his findings to experiences of pain. The animals, so he maintained, explain the excruciating pain suffered by people tormented by diseases of the bones. The accounts of sufferers—"it is as if the pain is deep inside the bones"—in Ruysch's view, reflected the pain one would feel if an animal was forcing its way inside. Using very visual language, thus stirring the imagination of his readers, he wrote that

> Since these remains of animals were found in the innermost recesses of the bones, it is permissible to believe that such pains are caused by little bloodless animals present inside the bones, which through *biting, gnawing and drilling* eat through the very thin membrane that surrounds the bone marrow (Ruysch 1744, vol. 3, 107, my italics).[14]

In order to justify his conclusion that pains in bones are caused by little animals, Ruysch further argued that sufferers usually felt better once they had been treated with quicksilver, which, at the time, was considered an effective antidote to worms inside the body.[15]

The story shows how closely visual evidence was linked to imagination and experience. It was only after a "very bright artist" had *shown* the animal remains (or whatever he thought looked like animal remains) to Ruysch that he himself began to notice animals inside bones all the time. Moreover, he linked what he saw to what he knew about the natural history of worms and flies. Additionally, the particular experience of pain in bones and joints made the presence of these animals all the more likely. Ruysch, as it were, visually represented the likelihood of worms and other bloodless animals gnawing their way inside.

Although Ruysch's discovery is puzzling to a modern reader, it was not so in his own time. The image of pain and discomfort in the shape of an animal or other monster inside was widespread. Not only did pile worms threaten the foundations of Dutch society, but contemporary satirical depictions and other illustrations represented pain as animals and little devils attacking the human body, images that in Ruysch's work became other than metaphorical.[16] Considering that worms were

---

[14] Ibid., vol. 3, p. 107, my italics.

[15] For instance, in his aphorisms concerning children's diseases Herman Boerhaave recommended medicines prepared with quicksilver against worms. (Boerhaave 1741, aphorism 1369).

[16] For a good collection of satirical images see Stafford (1991).

an ominous threat it is not surprising that "animals inside" were considered a very genuine threat.

Anatomists were considered the most suitable experts to investigate the animals. Pamela Smith has recently argued that in the early modern Dutch Republic "knowledge was acquired through bodily engagement with nature". Anatomists and surgeons, particularly, claimed to have more direct and authentic access to certain knowledge about the body precisely because they literally experienced the material substance of the human body (Smith 2004, 156–157). In the same way, our stories reveal that where physiological (*i.e.* theoretical) explanations ended, anatomy took over when it came to elucidating feelings of illness and distress. The animals *embodied* experiences of pain and thereby helped people to visualise it, and deal with it. Ruysch's findings, in other words, reconciled medical ideas *about* the body and the lived experience *of* the body.[17]

Secrets of Women

It is striking that the "Bosom Serpent" mostly affected female bodies. For this reason, a second aspect of animals inside is closely related to culturally determined ideas on gender and generation. Admittedly stories about animals in male bodies also circulated, but they were explained in different terms. Men, in almost all cases, suffered from animals creeping into their body from the *outside*, while animals in female bodies always seemed to be generated *within*. As a consequence, I suggest, stories about animals inside male bodies imply helplessness or innocence, as if men could not help it.

Ruysch, for instance, relates the story of a "distinguished" gentleman (whose name he did not mention). The man was bothered by an itch in his genitals, and his urine contained small elongated bodies which Ruysch put in a small box to take home. When he opened the box the next morning a bunch of flies flew away, leaving behind what appeared

[17] Traditionally historians of medicine have focused on ideas about the body; it is only since the 1980s that they have supplemented their research with experience of the body. Roy Porter has named this the 'patient's point of view'. (Porter 1985). Although Ruysch's case history of Grietje Willems does not reveal her own point of view, the little dog is a clear representation of a particular bodily experience. For the idea of the embodiment of knowledge see Lawrence (1998); Daston and Galison, (1992; Daston (ed.) (2004); Maclachlan (2004).

to be pupae. Ruysch explained that the man must have spent too much time on the privy, thereby enabling worms to creep up the urethra to the bladder (Ruysch 1744, vol. 2, 517–518).[18] Although Ruysch did not wish to mention the man's name, no shame is implied in the man's story. Ruysch simply reduced the gentleman's problem to imprudent behaviour and the consequent invasion of insects into his urethra.

In the case of women, however, the cases are almost always related to unknown processes of generation. The little dog vomited by Grietje Willems was a creature bred in the body—it was contained in a pocket resembling a womb, and it had an umbilical cord. The slug thrown up by the fourteen-year-old girl, also allegedly had an umbilical cord. Even the fact that Ruysch, of all medical men at the time, was called to the scenes corroborates my point. Who other than the city obstetrician and supervisor of midwives of Amsterdam, renowned for his skilled use of obstetric instruments, would be suited to judge such cases?

The stories of Grietje Willems and the fourteen-year-old girl, and particularly the fact that they had to be confirmed and thus legitimised by eminent men underscores the point recently made by Katharine Park: that female bodies were experienced as strange and uncontrollable. This was closely related to the female ability to generate independent life and the consequent association of the female body with complex *internal* anatomy. Knowledge about the female interior, in particular its sexual and reproductive functions' had become known as the "secrets of women" (*secreta mulierum*).[19] This concept was central to much of the work of male medical writers, as it was thought that

> women's bodies required intensive scrutiny and study, not because women were modest or malicious but because their bodies were intrinsically opaque, since, to their own disadvantage, the reproductive processes that were increasingly thought to define them took place deep inside' (Park 2006, 103).

---

[18] Ruysch (1744) [see note 3], vol. 2, pp. 517–518.

[19] The 'secrets of women' referred to the private access of women to knowledge about sexuality and generation that men did not have, as well as to the inaccessibility of information to men and women alike, simply because the process of generation is hidden inside the body. The term is derived from the Latin treatise *De secretis mulierum* attributed to Albertus Magnus, but probably composed by a German follower. The work circulated until well into the early modern period (Park 2006, 26 and 82). For a history of the idea of 'secrets of women', Park refers to Green (2000, 5–39). See also: Sherwood-Smith (1997); Sherwood-Smith (1999).

The case histories of Grietje Willems and others reflect anxieties about the generative force of female bodies and the anticipated possibility of bringing forth monsters and other abnormalities.[20]

For instance, there was a general fear (rumoured by midwives) of women bearing what was known as a "sucker" (*molla* or *zuyger*), a lump of flesh without bones, generated through an imperfect or evil conception. Sometimes these lumps contained hair, teeth or what looked like a limb, and often they looked like animals. Women even testified that they had felt movements inside, which reinforced the suggestion that they were carrying a live animal.[21] Stories even circulated of "suckers" running or flying through the room after birth and forcing their way into the vaginas of other women present. Ruysch was often presented with examples of such "suckers". In all cases, however, he showed that they were no more than lumps of clotted blood or the remains of the placenta.[22]

Yet, paradoxically, although Ruysch was keen to find a medical explanation for the phenomenon of "suckers", the case of Grietje Willems shows that the secret female body to some extent remained beyond natural description. The widespread belief in "suckers" shows that the link between an abnormal "birth" and an evil conception was easily made. Animals found in vomit were always suspicious. Not only Ruysch, but also the *dominee* and the bailiff believed Grietje Willems. At the root of their belief must have been the assumption that all sorts of unspeakable things could be going on inside a female body. In his explanation of the animal, Ruysch argued that it was generated by some physiological process which he had to admit, he did not quite understand. By exhibiting the little dog in his anatomical cabinets, however, he visualised and validated the generative force of the female interior and the possibility of women bearing animals.

---

[20] Ruysch referred to the work of Bartholin on unusual births (the same Bartholin who also investigated the above mentioned case of Catharina Geisslerin).

[21] Ruysch's son Hendrik, who was also an anatomist, explained that this could not be true and that what the women felt could be attributed to wind or unusual movements of the muscles of the stomach (Kooijmans 2004, 338–339).

[22] Ruysch's mission to explain 'suckers' as 'normal' products of the female body exemplifies the medical approach towards preternatural phenomena adopted by many Dutch doctors (Swan 2005, 157).

Conclusion

The stories of Grietje Willems and the fourteen-year-old girl show that the perception of the inner body is historical, embedded in a cultural context in which it has a specific identity and meaning. It could be argued that the little dog and the slug were hushed up abortions (of Grietje Willems' granddaughter and the fourteen-year-old girl). However, giving a modern explanation of the "animals inside" is not the point here. The stories about animals living inside show how such animals *literally* and *tangibly* illustrate how people experienced and imagined the workings of the interior body. For this reason the accounts have to be studied in their historical and cultural context. They show that a "historicization of the body does not limit itself to representations or bodily practices (...) but expands to include sensations and bodily experiences" (Pilloud, Louis Courvoisier 2003, 45).[23] The creatures in some way mediated anxieties about the interior body—a process which is not entirely alien even to today's thinking. Who, for instance, likes the idea of a tapeworm living inside them, and who did not flinch when watching Ron Weasley coughing up slugs in the film version of J.K. Rowling's *Harry Potter and the Chamber of Secrets*.

The alleged living and breeding of all kinds of animals inside the human body dovetails seamlessly with the early-modern Dutch obsession with little animals gnawing at dikes and houses and thus at the wealth and prosperity of the Dutch nation. In this way metaphors about the body of the Republic also work in the opposite direction—worms endangering the welfare of the nation were similarly thought to threaten the well-being of people. Just as pile worms drilled their way into the wooden structures of dikes and houses, "animals inside" attacked the bones and other structures which were important for the functioning of the human body. Although I have found no evidence of "animals inside" being considered a divine punishment, they still echo a general worry about physical as well as moral corruption. Why else would bystanders call the *dominee* and the bailiff?

Exhibits in Ruysch's cabinets were meant to remind the beholder of the fragility of life. The little dog and the slug were no exception. They served as tangible reminders of how easily animals can invade the

[23] For the relation between images of the body and the historical and cultural circumstances that shaped them see also Kuriyama (1999).

human body and cause excruciating pain and discomfort. Moreover, as pile worms loomed large in the public imagination, endangering the safety of the Republic, "animals inside" were extremely powerful symbols of corruption. This meant that visitors to Ruysch's collections could immediately recognise and almost "feel" the devastating effects of the animals. They could, as it were, experience the drilling of worms at the foundations of society as well as inside their own bodies. And no other experience could be more effective in the specific Dutch advocacy of individual virtue as the foundation for the survival of the Republic.

CHAPTER THREE

DEPICTING SKIN: MICROSCOPY AND THE VISUAL ARTICULATION OF SKIN INTERIOR 1820–1850

Mieneke te Hennepe

Skin is special. Nowadays considered to be the largest human organ, the skin occupies a remarkable position as mediator between the inner body and the outside world. Visible as bodily surface and boundary, physicians have used the skin in the nineteenth century to inquire into the health of patients, while microscopists examined it as an object to learn about physiology and anatomy. Only since the early nineteenth century has the skin been treated as an organ that mattered. In microscopical practices in Germany between 1820 and 1850 anatomists and physiologists increasingly produced drawings of the intimate anatomical structure of healthy and diseased skin. Inner structures of the skin itself were named, renamed, defined, redrawn and functionally redefined. A new image of the skin materialized in microscopic pictures in a process of *visual articulation*.[1] This paper argues that microscopic depiction of the skin reinforced a substantial shift in the meaning of the skin from an open porous cover towards a thick, functional and protective boundary organ of the human body.

In recent work on the history of microscopy historians of science have noted that the renewed interest in microscopy from the 1830s onwards was related to a context of new theoretical and social frameworks and not necessarily to the technological enhancements of the compound achromatic microscope (La Berge 1999; Ruestow 1996; Wilson 1995). In microscopical communities in Germany and Paris medical research on organs and tissues was growing. Cell theory was launched in 1839 and in the 1850s Rudolf Virchow introduced cellular pathology (Bracegirdle

[1] The term 'visual articulation' does not imply the visualisation of a pre-existing idea. Instead, I introduce the term 'visual articulation' to refer to the way the meaning of skin came into being *through* its depictions in the context of microscopical investigations between 1820 and 1850. Cf. Bruno Latour's concept of 'articulation' in Prentice 2005.

1993). In the climate of newly emerging microscopic communities comparative studies were also made of the structure of human and animal skin. Why this new interest in the skin? How were pictures involved in these new researches? What did the findings and pictures imply for the scientific and medical understanding of the skin and the body? In this paper the depiction of skin is taken as an activity deeply rooted in microscopical research. Not only did microscopists create pictures of the skin, the pictures were also actively involved in the construction of new ideas about the skin and the body.

Scholars of art history and cultural studies have linked the history of the skin to a profound shift in the conception of the human body around the time of the French Revolution. This shift is often connected to Mikhail Bakhtin's argumentation of a cultural change in perception of the body between the Renaissance and the onset of bourgeois modernity (Bakhtin 1984). The process entailed the replacement of a 'grotesque' image of the body as open and porous to a closed-off, delimited and individuated body. In her work on the body in the French Revolution historian Dorinda Outram has linked this change to Norbert Elias' notion of *Homo clausus* (Outram 1989). Outram argued that the dominance of the closed, individuated body image was only completed with the new political culture of the French Revolution. Scholars writing on the historical conceptions of skin in art, literature and medicine have taken up this idea of a changing body image towards an individuated and 'sealed-off' body in the late Enlightenment. They argued for a concurrent changing conception of the skin from an open porous layer to a closed boundary limit for the body (Benthien 2002; Fend 2005). Art historian Mechtild Fend explicitly associates the introduction of the term 'boundary' in the work by Xavier Bichat with the emergence of the skin as limit of the body (Fend 2005, 314). Yet most cultural historical accounts of the body have bypassed the production of knowledge on the skin in microscopical practices. From the perspective of medical history too, the microscopical redefinition of the skin in the formative period for investigations of bodily tissues between 1800 and 1850 has been under explored. This paper shows that the shift of body image gained currency in the production of a microscopical picture of skin in the early nineteenth century. The visual articulation of the *interior of the skin* in microscopic pictures defined a new idea of the relationship between the inner body and the outer milieu.

Skin pores were a focal point in microscopical investigations of the skin. Almost invisible to the human eye, these little openings in the skin received a lot of attention from microscopists. This chapter illustrates the shift in the image of the skin by focusing on the case of the skin pores. Illustrated texts by physiologists and anatomists in Germany and France from 1820 to 1850 are used to discuss how the skin was visually redefined into a thick protective organ for the body in microscopical research processes. First I discuss views on the skin pores that validate the 'openness' of the skin in the late seventeenth and early eighteenth century. The subsequent sections then explore how illustrations in microscopical research on the skin in the early nineteenth century introduced a novel image of skin. Finally, I will argue that the depiction of skin and the shift towards a closed-off body with a thickened skin are intimately linked developments.

Open bodies and porous skin

On a hot day in the summer of 1693 the Dutch microscopist Antonie van Leeuwenhoek (1632–1723) aimed his microscope at the skin of the palms of his hands:

> I cleaned well, part of the Skin of my Hand, and by my Microscope, in a space not bigger than a Sand, I saw the Sweat issuing out at about Fifty places, which as they touched, joined together into one little Bubble. After drinking about a Quart of *French*-Wine over Night, I found myself a little out of order the next morning, at Dinner I drank a Pint and half more, and after about Two Hours, I drank Half a pint of Tea very hot, that I might throw my self into a Sweat; (...) I examined it (...). I made this Experiment, to see if any of the salt Particle to be found in my Sweat, were like those found in Wine (Van Leeuwenhoek 1693, 954–955).

Van Leeuwenhoek's description of the microscopical inspection of his sweat referred to the idea of the skin as an open passageway for bodily fluids and substances. More specifically, Van Leeuwenhoek assumed that sweat was transported through the skin by means of "pores" (ibid., 955).

Apart from Van Leeuwenhoek, other microscopists also made observations of the human skin in the late seventeenth century. In the context of mechanical philosophy microscopic activities flourished between the 1660s and 1690s. The well-known English botanist and microscopist Nehemiah Grew (1641–1712) had published his investigations of the skin of the hands and feet in a short piece in the *Philosophical Transactions* (Grew 1684). In his descriptions Grew indicated how the existence of pores in the skin was a general and unquestioned matter for physicians:

> By *Pores, Physicians* mean no more, than certain permeable spaces between parts of a Body. Wherefore, that there are *Pores* in the skin of every man's Body, is no more to be question'd, than whether Men do ever Sweat or Perspire (ibid., 566).[2]

According to Grew, the pores functioned as an open passage for the discharge of "perspirable" and "noxious" parts of the blood "which by the continual use of the *Hands* and *Feet*, are plentifully brought into them" (ibid., 567). This was why, Grew argued, the sweat of the feet with many people "is much more offensive than any other part of the Body" (idem). The pores were thought to form a continuous open communication pathway between the body and its environment.

Eighteenth century accounts of bodily experiences by patients confirm the openness of the body and the porous conception of the skin. In her work on the medical histories of women in the writings of the German physician Johannes Storch (1681–1751) body historian Barbara Duden points to the conception of the skin as an open system (Duden 1991). In accordance with the humoural body, the skin pores were seen as bodily orifices and points of entrance or exit, like the eyes, nose, breasts, anus and navel (ibid., 120). The skin was "made permeable from the inside: it has 'sweat holes' which heat could open to allow the discharge of humidity, bloody matter, and impurities" (ibid., 121). Similarly, patients of the Swiss doctor Samuel Tissot (1728–1797) referred in their accounts to their evacuations from the bodily interior to the exterior (Pilloud and Louis-Courvoisier 2003). Medical historians Pilloud and Louis-Courvoisier have used the term 'porosity' in their characterisation of the body in the accounts of Tissot's patients implying "an exchange between interior and exterior of the body" (ibid., 470). Yet the openness of the body to the world in these accounts should not be delimited to the corporeal level: "The body is submitted to a logic that is both physiological and moral, maintaining its integrity only through a fluidity of exchanges between interiority and exteriority" (ibid., 472). Body and environment were both open to each other, with the skin pores as one of the important orifices for exchange.

---

[2] Italics in original.

By the early decades of the nineteenth century the conceptualisation of the skin changed. Although microscopy had flourished in the seventeenth century, the instrument was relatively little used in the eighteenth century, at least not for scientific purposes. In the early nineteenth century, however, the tool was reintroduced for scientific analyses and used in a widespread investigation into the physiology and anatomy of the human body. Within a few decades the microscopical exploration of the anatomical structure of the skin had put an end to the skin as open cover of the body. The image of an open, porous skin was replaced by a new image. In the early nineteenth century the skin, with the help of the microscope, was visually articulated as a functionally active, thick *organ*.

A new object of investigation

In the 1820s and 1830s the skin became the object of important types of analyzing practices. At universities in the German states anatomists and physiologists showed particular interest in the skin pores and tactile sense. Examinations into these subjects reconstructed the skin within changing perspectives on physiology and anatomy. In the course of the nineteenth century physiology and anatomy separated as disciplines. Experimental physiology replaced old anatomy and old physiology as a dominant discipline "for the investigation and discussion of the phenomena of life in the human and the animal" (Cunningham 2002, 632). In France François Magendie introduced a physiology that took animal experiments as the starting point for the study of bodily functions (ibid.; see also Hagner 2003). Not the anatomical facts, but the outcomes of experiments served as insights into bodily processes. Between 1770 and 1820 changes in the institutional German academic structure went hand in hand with new ideas about knowledge acquisition of the body (Broman 1996). In the newly established German system the body became the subject of a thorough investigation. Anatomists and physiologists took part in a large-scale investigatory enterprise to disclose the functions and morphological structure of bodily tissues. In this context skin became an interesting object of study.

Sense physiologists now became primarily interested in the secreting and the tactile functions of the skin. Physiologists at German universities such as Johannes Müller (1801–1858) and Jan Evangelista Purkinje (1787–1869) conducted experiments to study sensual perception (vision, speech and hearing). Key to their new approach was the experimental study of sensations. In a study on the skin Purkinje proposed a so-called "Individual-physiology", with the empirical investigation of an individual organism or organ as the basic principle (Purkyne 1823; Purkyne et al. 1979, 9–10). Historian of science Jutta Schickore denoted this specific methodological approach in sense physiology as exploring "how experience allows the subject to acquire objective knowledge" (Schickore 2003, 581). Contrary to the seventeenth-century microscopists who studied the skin, Purkinje considered the skin as a physiological object of empirical investigation. His experimental analysis shows how the skin emerges as an object of investigation *through* the experience of the researcher. Tactile experiences such as dryness, moistness or rigidity of the skin surface were put alongside the different possible outward appearances, colours and feelings of pain or coldness. Purkinje furthermore defined the skin as a system with an organic conformation (Purkyne 1823, 39). Pores were important parts in the configuration of the epidermis. However, Purkinje referred to the ongoing controversy about the existence of skin pores at the time. The physiologist Alexander von Humboldt (1769–1859), for example, had searched in vain for the pores as described by Van Leeuwenhoek with a compound microscope using 312-fold magnification (Eichhorn 1826, 29). Purkinje explained that pores were present in the epidermis, but they were hidden and invisible to the naked or 'armed' eye because of the elasticity of the surface (Purkyne et al. 1979, 125). The idea of an open skin remained intact, but Purkinje's experimental and experiential investigations marked the onset towards another conception of skin.

In the years following Purkinje's work other researchers adopted a similar experimental attitude towards the skin. Excretion and tactility remained the main focus in these studies. Purkinje had conducted his analysis of the skin with a magnifying glass, as did two German physiologists after him. In 1826 and 1827 two papers on the excretions of the skin were published in the *Archiv für Anatomie und Physiologie*. The German physician Heinrich Eichhorn from Göttingen wrote a lengthy paper on the excretions "through" the skin in 1826 (Eichhorn 1826). A year later Ernst Heinrich Weber (1795–1878), professor of anatomy in Leipzig, published on his observations of the skin, the skin glands and hair (Weber 1827). Both articles exemplify the doubts about the character of skin pores and the controversy about excretions by the skin at this time. Ernst Weber's description of the pores and the skin shows particularly well how he handled the skin as an object of physiological investigation, focusing on empirical knowledge about form and function. In previous years several physiologists had tried to prove the existence of skin pores in experiments. Yet nobody had been able to find the pores as described and depicted by Grew in 1684 (Weber 1827, 198–199). Trying several physiological experiments Weber also failed to find the pores. How was it possible that sweat came out, while the skin apparently was not a porous layer as assumed so far?

The porous nature of the skin became a contested issue in the early nineteenth century. The article by the German physician Heinrich Eichhorn emphasized the ambiguity of the position of the sweat pores. He claimed that the different secretions, such as sweat and sebum (skin smear), were excreted through sweat pores and by sebum producing glands of the hairs. After performing several observations, dissections and experiments, Eichhorn happily concluded that he had found the "sweat pores" with "sweat channels" of the skin: "Das Auffinden der Sweissporen mit ihren Canälchen hat mir in jeder Hinsicht ausserordentlich viel Vernügen gemacht; denn es ist damit gegangen, wie mit dem Ei des Columbus" (Eichhorn 1826, 475). The actual observation and description of his empirical experiments had been the only plausible way for Eichhorn to prove the existence of these "sweat pores", countering assumptions (ibid., 419). His account comprises over 50 pages of descriptions and instructions for observation and careful preparation and treatment of the skin as research object. Only with the right methods would the small sweat transporting channels become visible. The existence of the sweat pores as claimed by Eichhorn became entwined with an empirical-physiological method and context. These studies mark a turning point in the assumptions about the open porous nature of the skin. In the next decade microscopical studies into the anatomy of the skin would build on physiological investigations and the interest in skin as an object of investigation. Microscopic studies further changed the ideas about the skin pores in a crucial visual manner.

Visual articulation of skin as a mediating organ

A new picture of the skin was constructed in the 1830s. By then microscopists in Germany and France (Paris) were especially interested in the skin structures involved in excretion. In contrast to the previous physiological studies of the skin, the microscopic studies of the intimate composition of the skin included multiple illustrations. These illustrations mattered. They visually exposed the interior of the skin and its 'newly' discovered elements in schematic images. Now the "mysteries of the organisation" of the skin were the focus of investigations.[3]

A renewed interest in microscopy emerged in German universities from the 1830s. In Breslau Purkinje—holding the chair of physiology—introduced his students to the Plössel microscope, a new and powerful achromatic instrument (Coleman 1988, 24). He obtained the microscope in 1832, which marked the onset of a new period in his research (Kruta 1975, 216). With some of his students Purkinje started a collective research project devoted to the study of the microscopic anatomy of human tissues, what would later become human histology (Coleman 1988, 21). Microanatomists also aimed their microscopes to the skin tissues. One of Purkinje's students, Alphons Wendt reported on his microscopical observations on the structure of the epidermis, the outermost layer of the skin, in a paper in 1834 (Wendt 1834). Among his observations, one of most spectacular descriptions concerned the microscopical observation of the "sweat channel" as a neatly delimited "organ" in the skin (ibid., 285). Contrary to existing ideas about the secretion of sweat, which assumed the transportation of sweat *through* the skin, Wendt found that the sweat was produced inside the skin itself. The concept of the pores as closed holes was an important claim that indicated a transformation in the image of the skin. The skin no longer seemed an open passageway for transportation, but more a mediating organ harbouring functionally organized structures and processes of its own.

The microscopical observations of the new structure inside the skin entailed a particular visual specification as well. The illustrations of the sweat "Fäden" functioned as a visual articulation of the newly found structures. Wendt's article contained six illustrations. One figure of the engraved plate showed the different forms of the Fäden, the course and the structure of the observed elements inside the skin. In a very schematic way the illustration reinforced the idea of the structure as a little demarcated organ ending in the skin.[4] Wendt depicted not a channel through the skin, but an enclosed sack ending within the skin. Again, the new structure needed specific instrumental and method-

[3] The French anatomists Breschet and de Vauzème referred to "les points les plus mystérieux de l'organisation" of the skin (Breschet and de Vauzème 1835, 3).

[4] Wendt used the terms "Ende", "genau begrenzte Organe" and "geslossenem" when referring to the ways of the sweat in the skin (Wendt 1834, 285).

ological circumstances to become visible under the microscope. Special controlled preparation methods and instrumental preconditions were necessary for visualisation and depiction. Jutta Schickore has referred to the importance of these quantitative physical and chemical methods for microanatomists in the 1830s to gain control over their objects and render them more complex (Schickore 2001, 169). Visualisation of the inner structures of the skin just as well gained a greater experimental conditioning of the skin as an object of microscopical study.

Hundreds of miles from Breslau anatomists in Paris were also interested in the microscopic structure of the skin. In 1834 Gilbert Breschet and Augustin Roussel de Vauzème published their new investigations into the structure of the skin in the *Annales des Sciences naturelles* and as a separate publication one year later (Breschet and de Vauzème 1834; ibid., 1835). The main author, Gilbert Breschet (1784–1845), was a surgeon at the Hôtel Dieu and professor of anatomy at the Faculté de Médecine in Paris. With their study Breschet and de Vauzème aimed to contribute to the little known structure of the skin so far with empirical microscopical observation. According to them, previous research on the skin was not accurate and lacked information on the "organic elements of the cutaneous tissue" (Breschet and de Vauzème 1835, 1). Breschet wanted to correct the lack of knowledge by repeating observations with the microscope and thus uncover the "intimate composition" of the skin (ibid., 2). Depictions of these intimate elements were important components of the new ways of analysing skin. Breschet and de Vauzème included schematic black and white copper engraved figures of the 'internal' organs and structures composing the skin, with the sweat producing organ as one of the main components. Visually, the sweat organ was portrayed as a little hairy ball, with a spiralled tube opening on the skin surface. In his explanation of the figures Breschet described the appearance of the morphology of the sweat excreting system when seen with a strong magnifying glass and a microscope. A visual image of the internal structure of the skin was brought into being.

However, the intimate structure of the skin did not betray its appearance very easily. Because of the complex structure skin was extremely difficult to prepare. The dense, elastic and resistant quality of the skin made it almost impossible to study and depict (Breschet and de Vauzème 1835, 6–7). Analyses of the skin had so far been represented in pictures of square specimens of skin without exposing its internal structure. In his study with de Vauzème Breschet solved this problem by reverting to depictions of *schematic* compositions of the structures inside the skin. Just like the study of the skin part by part, Breschet depicted the separate elements or functional organs and then composed a schematic figure of the skin. Without schematization the entire composition of the countless quantities of structures in the skin would compromise the clarity of the relationships between the different structures. Figure 3.1 is an example of a depiction that illustrates how the skin was visually thickened in these schematic compositions of the microscopic structures inside the skin (see fig. 3.1). This picture showed the "composition of a synthesized figure or *Scheme* of the human skin" (Breschet and de Vauzème 1835, 6–7). Each part of the skin in this figure symbolized a different functional structure. The production of sweat for example was articulated by the sweat producing organ and its channels to the surface ending in the pores.

Within the context of Paris medicine Breschet belonged to one of the earliest generations of physicians using the microscope (La Berge 1994, ibid., 1998). What's more, Breschet's work on the structure of the skin connected Parisian ways of analysing skin with German ways of analysing tissues in a significant way. Whereas in Germany physiologists stressed the generative and developing nature of human structures and forms, Parisian physicians emphasized dissection and morphologies of normal and abnormal tissues.[5] With his accent on the structural organisation of the skin and the comparison between anatomical forms of skin, as well as his microscopical interest in anatomy and physiology, Breschet connected both ways of analysing. He recognized that without a proper knowledge of the anatomy of the skin, insights into physiological functions of the skin would remain obscure (Breschet and de Vauzème 1835, 2).

It did not take long before Breschet's findings and pictures of the structures inside the skin were picked up by microanatomists working on skin in German settings. In 1835 Ernst Friedrich Gurlt (1794–1882) published a paper on comparative research of the human skin and the skin of pet animals (Gurlt 1835). He argued that only after the latest developments in anatomy and physiology had Breschet and Roussel de Vauzème finally proven the true existence of the "sweat glands" (ibid.,

[5] On different forms of analysis in science, technology and medicine see John Pickstone's *Ways of knowing: a new history of science, technology and medicine*, in particular chapter 5 (Pickstone 2000).

Fig. 36.

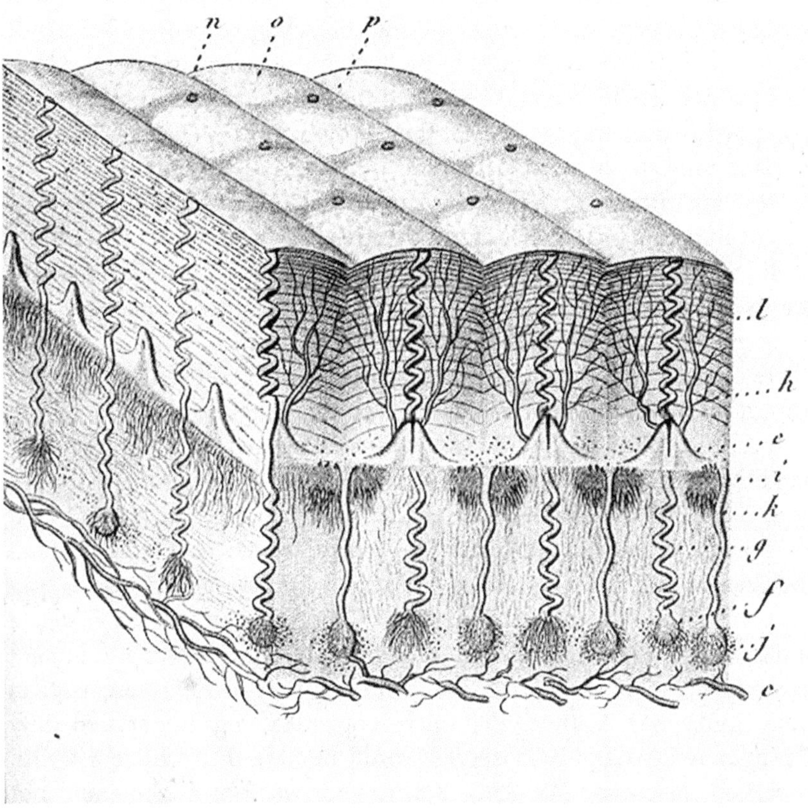

Fig. 3.1. Schematic microscopical articulation of the inner structures inside the skin in Breschet and de Vauzème's *Nouvelles recherches sur la structure de la peau* (1835). Universiteitsbibliotheek Leiden.

402). All the work before the studies of Breschet was dismissed. Again pictures played a pivotal role in the argument. When discussing the sweat glands, for example, Gurlt compared his own findings in cross-sections of human, horse, cow, dog, swine and sheep skin with Breschet's illustrations. He commented on the size, form and position of the sweat glands as they appeared in Breschet's figures. He further argued that Breschet and his colleague did not properly explain the typical structure of the little balls in the sweat glands and that this structure was not well indicated in their illustrations. Taking the schematic compositions of the internal structure of the skin by Breschet as a starting point, Gurlt thus further articulated and redefined the specifics of the sweat glands in his article and engravings. He thereby continued the schematic way of visualising the inner structures of human (and animal) skin.

Drawing microscopic objects formed a crucial part in the entire process of microscopical observation. In manuals on microscopy authors explicitly discussed the drawing of microscopic objects by referring to the use of devices like *Sömmerring's mirror* and the *Camera Lucida* for example (Vogel 1841). Instruments were used to secure a proper microscopical drawing, but additional artistic skills were still considered important. Drawing devices helped to sketch the general outline of the picture, but the details of the drawing still needed to be filled out by hand (Mohl 1846, 21). Microscopists further emphasized the importance of drawing for the entire experience of microscopical observation. In an essay on the microscope in the *Handwörterbuch der Physiologie* Jan Evangelista Purkinje argued for the need of drawing when observing with the microscope (Purkyne 1844, 433). The aim was to visualize and externalize the acquired subjective "inner" experiences and images of the microscopists on "outside" paper as part of the microscopical work (ibid., 434). As the Dutch microscopist Harting later argued, only a proper microscopical observer would be able to produce a useful depiction (Harting 1848, 336). Yet, since it was not possible and not necessary to depict all the details of the field of view, it would suffice to be selective and reduce the drawing to the characteristic and essential elements. Only in very few cases everything could be depicted. As contradictory as it seemed, Harting stated how a faithful depiction was not always a completely true depiction (Harting 1848, 338). Microscopists thus underlined in general the necessity of depictions as concise reflections of the experience of microscopical analysis of a particular object and as (schematic) articulation of the essentials of the object. Research and depiction went hand in hand (Purkyne 1844, 435).

The microscopical depictions articulated a 'thicker' skin as it came to harbour more and more inner structures. The flat, layered and porous skin was replaced by a picture of an organ that harboured multiple active structures, including the secretion of sweat. In the context of microscopical analysis this symbolical scheme of the inner make-up of the skin was brought into existence by different anatomists, in different settings and in different sequences of figures. A common scheme of the microscopic skin as it developed during the 1830s and 1840s, however, remained a standard for many decades, in many contexts. Eventually, Gurlt's figures served as a template in the standardisation of a microscopical depiction of the sweat glands of the skin. His were seen as "the first faithful depictions of the actual glands" (Krause 1844, 127) and thus appeared as copies in other books (Wagner 1839). A standardized depiction of the microscopic structure of the skin, including a hair, sebaceous glands and sweat glands, was shown in anatomical works and many books on skin diseases from the middle of the nineteenth century (see fig. 3.2).[6] Although initially the microscopical pictures were mostly shared and distributed among specialists and professionals such as dermatologists, anatomists and microscopists, microscopic pictures of the skin were eventually also printed in books for other audiences. For example, a depiction of a cross-section of human skin enlarged by the microscope appeared with articles on skin in the popular French *Larousse* encyclopaedia in 1897 and in the German *Brockhaus* encyclopaedia of 1898 (Augé 1897; Brockhaus 1898). For the general public too the pores in the skin had been visually transformed and articulated as part of a structure inside a thickened skin.

Thickening skin

Microscopical depictions of skin were becoming more and more part of bodily anatomy. In microscopical anatomy the thickened image of skin with the sweat glands as dominant structures became widely accepted. Influential textbooks on histology in the 1840s and 1850s included the

[6] Many atlases and handbooks on skin diseases and bodily anatomy opened with a section on skin. See for example *Die Hautkrankheiten durch anatomische Untersuchungen erläutert* (Simon 1848, Taf. I), *Mikroskopische Anatomie oder Gewebelehre des Menschen 2.Band 1.Hälfte: Specielle Gewebelhre von der Haut, den Muskeln, Knochen und Nerven* (von Kölliker 1850, 1), *Lehrbuch der Hautkrankheiten* (Neumann 1869, 10).

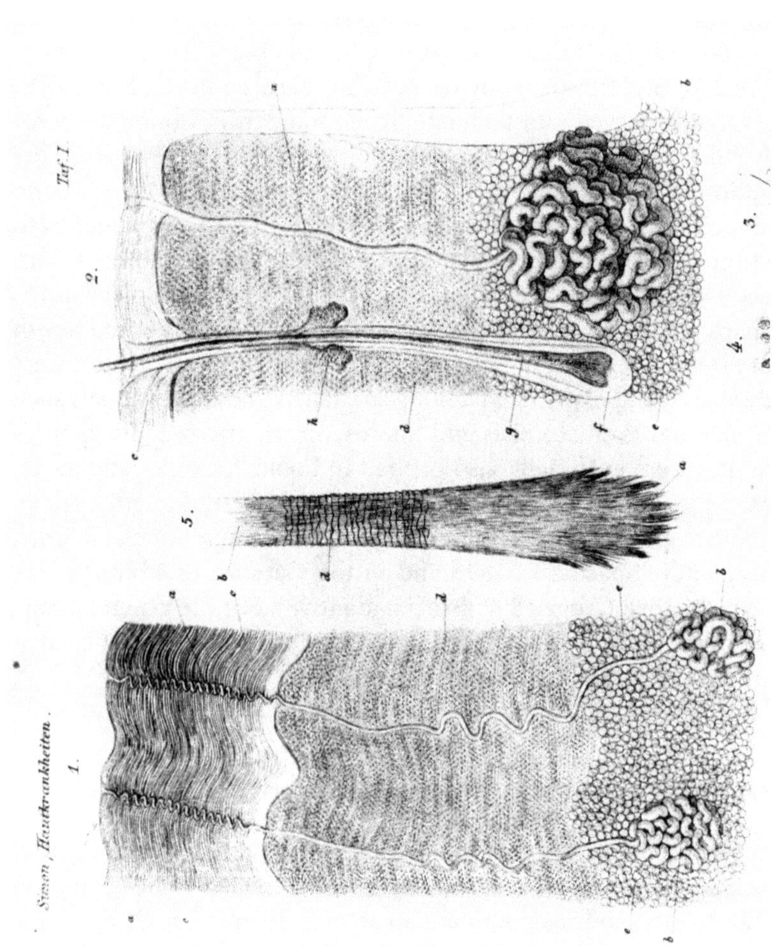

Fig. 3.2. Standard microscopical depiction of the skin in the influential histology handbook by Albert von Kolliker (1852, 1). Museum Boerhaave.

standard microscopical picture of the skin since in general pictures had become "almost inevitably necessary" in this field (Henle 1841, viii). Moreover, the construction of a microscopical image of skin signalled an important step in the shift towards a conception of the human body as a closed-off entity. In many medical and cultural histories of bodies the skin is discussed as an important surface in new hygiene discourses in the nineteenth century (Sarasin 2001, 264–313). The visual redefinition of the skin from a permeable layer into a mediating organ with its own internal organelles was an essential part of the emergence of the body as an internal milieu. The important conception of the human body as an organism with an 'internal milieu' was put forward by the famous French physiologist Claude Bernard in 1857 (Bernard quoted in Corbin et al. 2005, 39). His programme placed the scientific activity of experimental physiology at the heart of clinical medicine. The skin was redefined as bodily object for medical science. Medical understandings of skin cleanliness during the nineteenth century further promoted the importance of healthy skin in society. Pores blocked with filth and dirt would obstruct the healthy function of the skin as "sewage" of the body (Chadwick 1877). A redefinition of the skin thus involved broader changes in definitions of body and health in science and medicine.

The emergence of a microscopical image of skin allowed for a new idea of bodily interiority. Important cleanliness regimes from the nineteenth century onwards heavily leaned on the new conception of skin. More generally significant is the way in which the skin became a barometer for bodily health and hygiene. Management of the skin was taken up as a way to control the body. By visually exposing the interior of the skin, microscopists in the early nineteenth century thus not merely generated a different image of the skin but a different concept of bodily boundaries. Indeed, a shift in the concept of the skin entailed a new attitude towards the interior of the human body. This could only happen in concurrence with visual articulations of excreting organs inside the skin such as the sweat glands. Visually exposed, the two millimetre layer surrounding our bodies now counted as an organ that mattered.

CHAPTER FOUR

THE MIND AT WORK: THE VISUAL REPRESENTATION OF CEREBRAL PROCESSES

Michael Hagner

Images of the brain

The debate surrounding cerebral representation and the localisation of mental processes has become lively of late. Not that the discussion of the relation between the brain and the mind, which has been waxing and waning for a long time, has been enriched by any particularly original theoretical insights. Rather, the insights are visual, based on a number of imaging techniques that go beyond the scope of any visualisation techniques known so far in the neurosciences. In the history of the human sciences, innovations in visualisation technology have affected the perception of the physical and mental nature of man. Neuroimaging has transposed psychological phenomena into visual categories and thus changed their epistemic and cultural status. How profound and sustained these changes will be is not yet clear, but basically two scenarios are possible. Either images of the brain will come to play a fundamental role in determining our identity, at least in some areas. Or they will eventually lose their attractiveness, it being unproductive in the long run to contemplate pictures associating presumed mental states with certain regions of the brain if they fail to provide any kind of deeper understanding of how brain states and mental states relate. Even if I tend to assume that the fascination of these pictures will prove to have its limits, the subject of this article is to discuss possible consequences of an anthropology based on images of the brain.[1]

[1] An earlier German version of this article was published in *Anatomien medizinischen Wissens*, C. Borck (ed.), 259–286. Frankfurt a. M.: S. Fischer 1996. In the mean time I have somewhat revised my opinion. When I wrote the first version of this text in 1995 under the impact of the technical potential of neuroimaging, I was pretty certain that it would fundamentally change the way we see ourselves. Meanwhile, I am no longer so sure. The images are new, but the interpretations that are being offered to make

Traditional x-ray technology did not provide any insight into the functioning of the brain. Nor did computer tomography, which was hailed as revolutionary when it was developed almost 40 years ago and can now probably be considered a transitional technology in the move from radiological to computer imaging. Although the development of computer tomography was closely linked to the dissatisfaction at the time with the radiological techniques for the representation of the brain (Blume 1992, 159–60), the hopes for it were limited to the domain of clinical medicine and hardly affected the cognitive neurosciences. With the new imaging techniques—fMRT (functional magnetic resonance tomography), PET (positron emission tomography) or SPECT (single photon emission tomography), to name only the most important—the situation has changed. Images of the brain produced by means of these techniques grace half or whole pages in glossy magazines and daily papers. They give the impression of opening a new window onto the functioning of the brain and hence also onto the mental life of man, an impression that captions and comments do their best to reinforce. Taken for itself, such rhetoric is not particularly noteworthy. In one way or the other it accompanies every technological innovation. The introduction of x-rays at the beginning of the twentieth century was also met with a great deal of rhetoric (Cartwright 1995; Dommann 2003). But the cultural perception and interpretation that has arisen in connection with neuroimaging is more than just a fashionable fascination with the new visualisation of cerebral processes. Assumptions, ideas and value judgements that have their origin in the nineteenth century and that, one would have thought, have long been laid to rest, are being mobilised to interpret these images. This revival is indicative of present-day anthropological sentiments and views, which the images may well be transmitting, but which are not reducible to the same.

For a clear understanding and assessment of these images, it is important that we place them within the larger context of another deep-rooted change that has affected image-making: the transition from ordinary photography, based on optical principles, to computer-supported images that are no longer based on the same. The technique known as BOLD-fMRT (BOLD=Blood-Oxygen-Level-Dependent) takes

sense of what they show are not. That is something I would not have expected at the time. Among the studies that have since been published on this subject, see Hagner 2002; 2004; 2006; Dumit 2004; Weigel 2004; Burri 2008.

advantage of the fact that deoxygenated haemoglobin in the blood vessels of the brain is magnetic, whereas oxygenated haemoglobin is not (Hüsing et al. 2006, 21–50). The application of a strong magnetic field permits the measurement of the respective distribution of hydrogen atoms. A complicated mathematical operation is required to transform the obtained data into an image. Imaging is an appropriate term for this procedure insofar as it conveys the fact that the image is not a simple depiction of the object but rather something that is produced. The data could just as well be transformed into a graph or a distribution curve. But the latter would not have nearly the visual impact of the colourful brain pictures, which give the somewhat misleading impression of immediate evidence.

The new images of the brain, then, are the result of a process of production, yet they refer to an organic reality. Measuring devices record activity and, following a number of transformation steps, physicians or neuroscientists interpret it. Like photography or radiology, the new imaging techniques require a body. That distinguishes them from simulated computer images. As Roland Barthes observed for photography, one cannot deny "that the thing was there" (Barthes 1981, 86). But what is it that is being measured? Basically just a minimal rise in blood flow in a particular region of the brain. More precisely: blood at one location contains somewhat more oxygen than at another. That is what one sees in the brain scan. This activity, which is first measured and then visualised, can be correlated with increased neuron activity. In places where nerve cells are firing more, more oxygen is needed. But here already a problem arises: when the neurons of a particular region are activated, it takes 3 to 8 seconds for this region to be provided with fresh oxygen-enriched blood. Initially, there is even a drop in oxygenated blood, precisely because it gets used up immediately by the nerve cells. Then, after a few seconds, there might even be more oxygenated blood available than the neurons need for their activity. That means that there is a correlation between blood flow and nerve activity, but with a time delay that to date is not exactly determinable. The two variables cannot be correlated exactly. It is likely that this problem will be better solved in the future by means of a combination of fMRT and direct measurement of neuron activity by means of electroencephalography, but it is not clear to what point it will be possible to correlate the relation between the two processes exactly. These considerations only concern physical processes in the brain. How the minimal rise in blood flow relates to mental activity is an open question.

In light of the above, the widespread metaphor which would have the new computer imaging machines virtually opening the head and watching the brain as it carries out its activities does not seem particularly helpful. The comparison suggests that we are simply enabled to see what we normally cannot, implying that the instrument itself has no significant effect on the relation between the observer and the observed, the measurer and the measured. Neuroimaging, according to this notion, is essentially a refined, exteriorised sense organ. But technical instruments have been notoriously underestimated as exteriorised sense organs since the nineteenth century. They are not in fact just silent laboratory assistants always at the service of the experimenter. Techniques of manipulation and representation transform things into epistemologically relevant objects; instruments and apparatus acquire their experimental significance in several stages of adaptation and adjustment as the scientists develop their procedure.

So the question arises: is there anything radically new about the present and future images of the brain? Or do the new imaging techniques differ from the older ones only in their degree of technical refinement? Is the imaging technology of the early twentyfirst century a continuation of the various technologies that have been developed since the nineteenth century, when registering devices redefined the organic systems of the body? Or can one discern a new neuro—and biopolitics that presumes to determine, by technological means, the essential nature of life, the mind or mankind? Is there a new alliance of techniques and power in the offing that promises to establish a whole new way of looking at brain function and at the mental life of healthy and unhealthy individuals? Certainly, no one will want to deny the medical and cultural authority of the new brain images. But what do they have to say about human nature? These are the questions I would like to consider. In the first part of this paper I will present a brief historical overview of the older images of the brain. In the second part I will distinguish the recent brain images from the old ones, and point to several possible consequences of this difference.

Morphological and functional visualisations of the brain

For the sake of a historical account of visual representations of the brain, I consider it useful to make a distinction between morphological and functional images of the brain. The purpose of the former is to capture and represent a material structure. Every anatomical representation, be it a drawing by an eighteenth century anatomist or a twentieth century photograph of a histological brain specimen, presupposes a natural point of reference in the represented organ. Anatomists were well aware from the beginning that things were not so simple: in order to create a suitable representation, a number of rules needed to be established and respected. In the eighteenth century, it was primarily a question of the appropriate preservation of the organ, so that it could be further sectioned, stored in a jar or reproduced in a more or less realistic drawing. Then, it was a matter of finding the right point of view for the representation, not that of the chance observer and sketcher, but the schooled view of the anatomist. Samuel Thomas Soemmerring, who as one of the most important anatomists of his time was famous for his extremely exact representations wrote:

> It is well-known that it is very difficult to find a drawer who on the one hand is capable of understanding everything that a dissector actually needs to know, and on the other hand is not too stubborn to follow a lead. Most of them apply to insignificant details, to an unnatural fold due to the shrinking of the specimen in the alcohol... a meticulousness that does harm to the important points. Some artists simply cannot be taught to see only what is actually supposed to be represented and to leave out the details that do not belong to the subject or are even merely accidental. (Soemmerring 1791, 5).

According to Soemmerring, the anatomist does not see the body as does the artist, because his interests and practiced way of viewing are different. The creation of an anatomical reality thus presupposes a disciplining of the view and of the drawing pen.

As representational techniques became more complex, the problem of accuracy and the danger of mistaking an artificial detail for a natural body feature became more acute. In microscopic anatomy, for instance, not only did the eye have to be disciplined, but the subject of study itself had to undergo various complicated procedures (Chadarevian 1994; Schickore 2007). The visualisation of cells with the aid of the microscope depended upon the methods of preparation and dying that were employed, each of which made different structures visible. But what did these methods actually do? Was the tissue still natural after being treated? If certain cell structures were represented and others not, could one get a reliable image of the nervous system? Problems such as these were prominent in late nineteenth and early twentieth century discussions about the status of nerve cells and the construction of the nervous system (Breidbach 1993, 109–115). Still, in spite of all the uncertainties that reigned at the time, one thing was beyond doubt: it only made sense to look at a nerve cell under the microscope because it was accepted as a constituent element of the brain. Morphological images of the brain always had a natural correlate.

Functional images of the brain are likewise the result of techniques, experiments, hypotheses and models. But their purpose is not to create a morphologically true likeness of a material structure. Rather they serve to explain functions, in the broadest sense of the term, or to indicate an association between functions and certain body structures. They are not re-presentations but rather sketches, diagrams or graphic records. Such images are generally regarded as a provisional means of clarification and therefore secondary to the actual research process. But to discount them as mere didactic aids or illustrations is to underestimate their truth claim. Functional images have a history of their own independent of morphological images, as becomes clear from the development of the cerebral localisation of mental faculties. It first became possible to visualise the brain otherwise than in morphological images when, in the nineteenth century, brain scientists began to associate certain areas of the brain with certain mental faculties. This meant that the latter became objects amenable to scientific study. How was this amenability translated into visual images? How did these images change as localisation research progressed? And how did they themselves become vehicles of knowledge implying a truth claim?

The visual representation of physical or psychological brain functions is as old as the idea of localising mental properties in the brain. Phrenological images in the tradition of Franz Joseph Gall provided the basis for the first functional maps of the brain. Gall programmatically introduced the idea of the cerebral localisation of mental tendencies, talents and properties into the human sciences at the beginning of the nineteenth century, eliciting considerable uneasiness at the time. The assumption of special organs in the brain for calculating, religiosity, charity or miserliness was understood by many contemporaries as a threat to the autonomy of the human subject. The popular practice of palpating the scull in order to draw up a personality profile was an object of mocking derision.

Figure 4.1 illustrates the miserable state of phrenology that was spurned by official science: all manner of mental properties and tendencies are entered into small bubbles on the scull. There are more organs here than Gall ever conceived of, but the essential thing to note is the

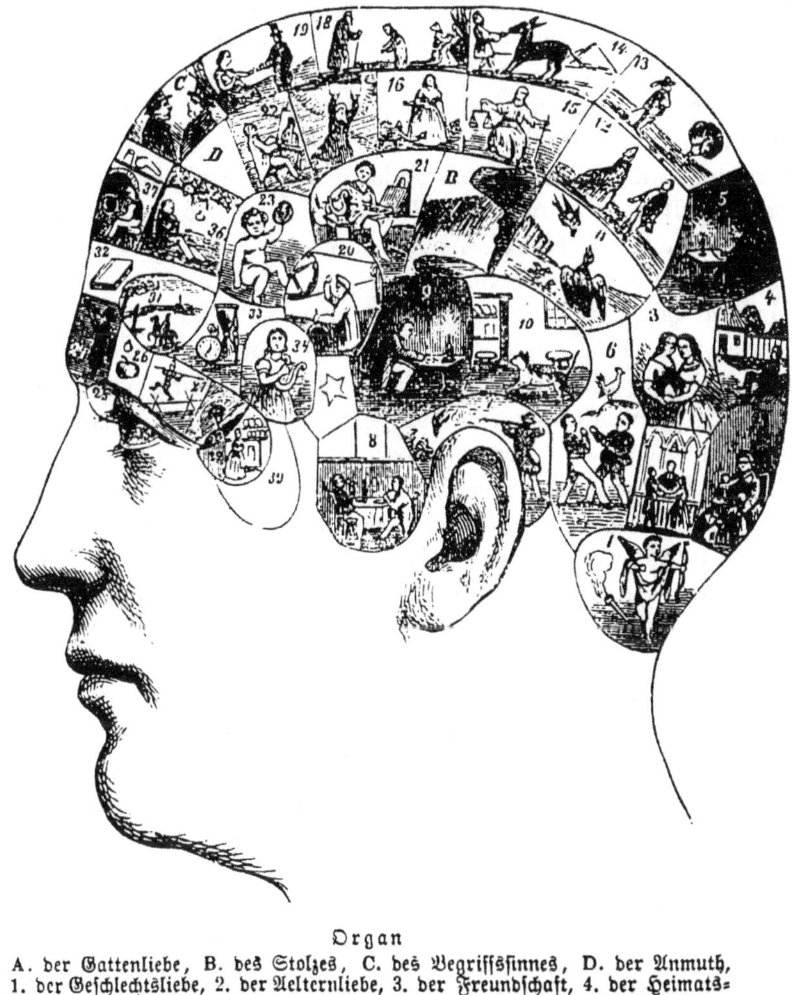

Fig. 4.1. An iconic phrenological map of the brain. Archive of the author.

iconographic figurines in the medieval and modern tradition of pictorial representation (fig. 4.2). Even the head, which these little pictures adorn like a bathing cap, recalls the heads of the medieval ventricular theory, in which the faculties (sensus communis, ratio, phantasia, memoria) were inscribed into cerebral bubbles. Although Gall intended something entirely different from the ventricular theory, the visual connection of phrenology with this tradition branded it as not belonging to scientific discourse. It thereby undermined its own claim of presenting scientifically tenable results. No science that took itself seriously needed to concern itself with images of this sort.

In spite of all the criticism of phrenology, Gall's basic assumptions acted as catalysts for the further development of brain research when anthropology, biology and medicine became the leading sciences of a new quantitative and physical study of man in the second half of the nineteenth century. A mesh of physiological, anatomical and clinical methods, techniques and devices were applied to study mental faculties, and it became ever more accepted to see the brain as a map onto which these faculties were successively entered. The functional brain images that served as a means of orientation in this regard were remarkably unspectacular—hardly more than schematic sketches at first, whose anatomical accuracy was only approximate. Hence the explosive power of the sketch shown in Figure 4.3, localising the brain's speech mechanism, was hardly discernible. In 1874, the psychiatrist Carl Wernicke did cause somewhat of a stir with his aphasia theory, which synthesised elements of anatomy, physiology and clinical pathology. To the frontal speech centre, in which Paul Broca had localised motor aspect of speech together with so-called "movement images", Wernicke added a posterior, sensory centre, which also included "memory images" of past sense impressions (Wernicke 1874, 19). This cerebral mechanism is depicted in the figure: x stands for the sensory speech centre, y for the motor speech centre, ax represents the auditory pathway, ym the speech output pathway. With this depiction of an interconnection of functions, Wernicke presented a model of the brain as a sensory-motor machine. The strength of his model lay in the fact that it was based on anatomical and clinical studies that seemed to suggest such a functional differentiation of the cerebral cortex. This provided the background for Wernicke's description of the speech mechanism as a "cortical reflex arc", in which impulses were received, transformed and sent out. Speech became, at least for the purposes of research strategy, no different from any other physical function, and that is exactly what the sketch was

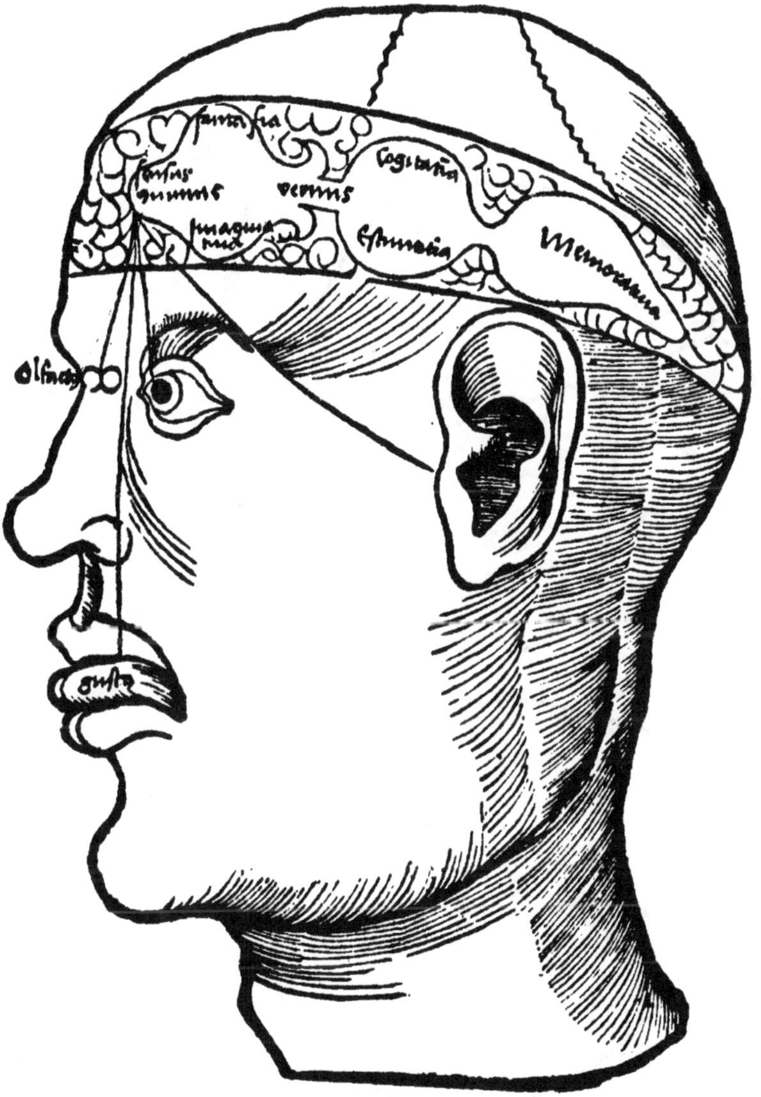

Fig. 4.2. A typical medieval map of the brain. Faculties like sensus communis, fantasia, aginativa, cogitativa, estimativa and memoria are localized in three ventricles (Gregor Reisch, Margarita philosophica. Basle 1517). Pietro Corsi (ed.): The Mill of Thought. From the Art of Memory to the Neurosciences, Milano 1989.

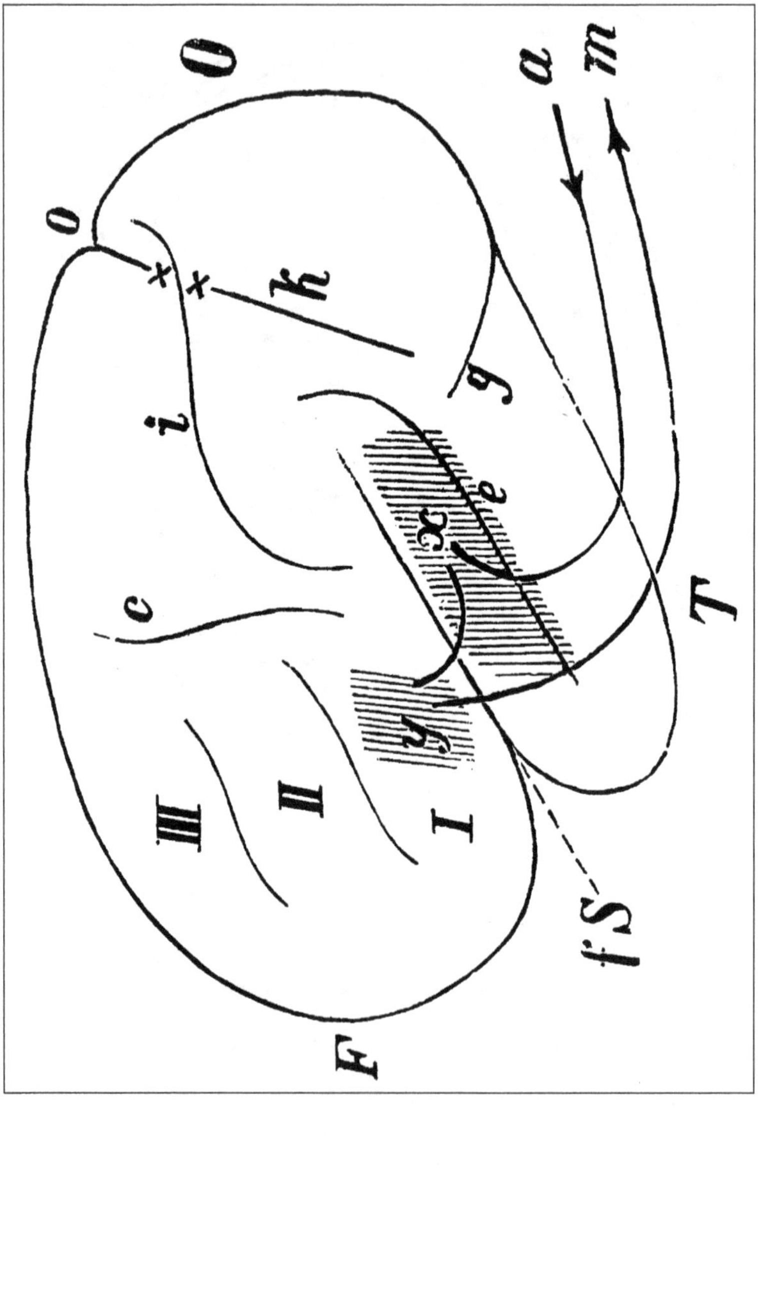

Fig. 4.3. Carl Wernicke's iconic representation of the cerebral mechanism of speech. For details see text. Carl Wernicke: Lehrbuch der Gehirnkrankheiten, vol. 1, Kassel 1881.

supposed to show. For this kind of visualisation, anatomical precision in the representation of the two speech centres was not required.

After Wernicke, there followed a process of diffusion, a systematic search for regions of the brain that could be functionally isolated. This search focused above all on properties that had previously been neglected, such as seeing, hearing or somatic and sensory functions. Initially their localisation was the subject of systematically planned series of experiments. The Berlin physiologist Hermann Munk particularly distinguished himself in this respect. His localisations in the cerebral cortex went beyond the topographic differentiation of the above-named functions (fig. 4.4). Just as Wernicke had made a distinction between a sensory and a motor aspect of language, Munk distinguished, in his experiments on the system of sight, between blindness and mental blindness ("Seelenblindheit"), and in his studies on the motor and sensory system, between the paralysis of particular body parts and mental paralysis ("Seelenlähmung"). Each displayed respectively different symptoms and followed upon the injury of respectively different brain regions. Mental blindness referred to the absence of visual memory of familiar objects; mental paralysis meant, in Munk's crude experimental language, that a monkey "who is mentally paralysed in a frontal extremity not longer knows how to bring his hand to his mouth, no matter how great his greed for the treat he holds in his hands" (Munk 1890, 59). The important thing to note here is that it was not the behaviour of the pitiable monkey that brought Munk's research onto its trail. On the contrary, it was the possibility of representing the anatomical reality of a damaged brain area that made the monkey's behaviour into a clearly definable deficient action. That is what Munk's figures were meant to show. They indicated the place of the experimental intervention, and were meant as a guide for further laboratory use.

Wernicke's and Munk's images were far removed from the exhaustive phrenological depictions of mental representations, even if the programmatic claims of the localisation theory did point in the direction of an exhaustive study of organic functions. The sparseness of their figures helped ward off not altogether far-fetched associations with Gall's organology. The same can be said of a drawing by the Swiss neuroscientist Constantin von Monakow, even if it no longer just marks the location of an experimental intervention or a pathological lesion. Monakow's aim was to identify motor areas in the human cerebral cortex (fig. 4.5), for which end he drew upon data from a wide variety of sources. What looks like a homogenous set of data on his picture was actually in part

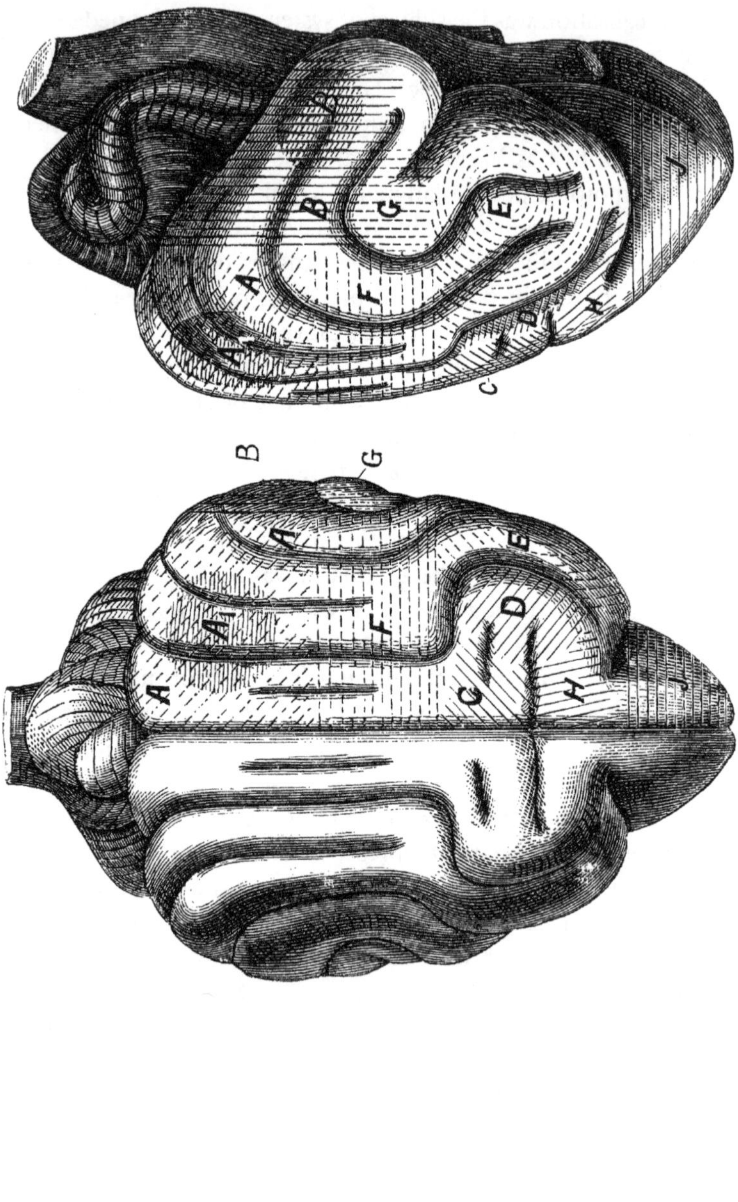

Fig. 4.4. Hermann Munk's depiction of the cerebral cortex of the dog: A is the optical sphere, B is the acustic sphere, C-J sensory spheres. Hermann Munk: Über die Functionen der Grosshirnrinde. Gesammelte Mitteilungen. 2nd ed. Berlin 1890.

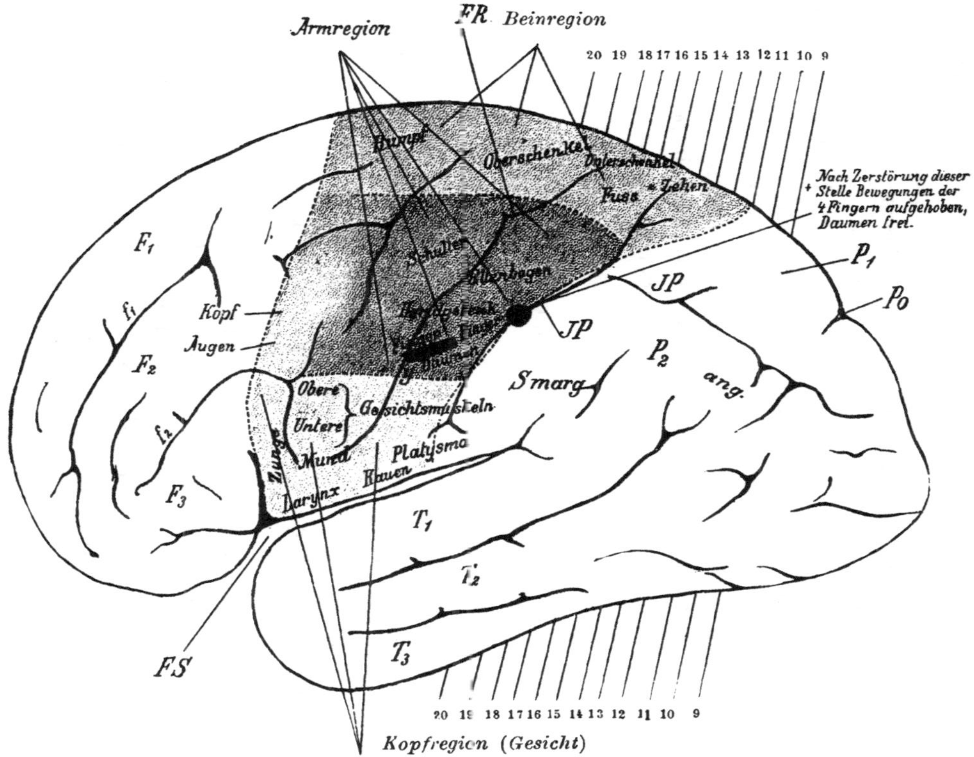

Fig. 4.5. C. v. Monakow's depiction of the cerebral cortex. Constantin von Monakow: Gehirnpathologie, 2nd ed., Vienna 1905.

derived from the results of animal experiments and part from observations of patients with brain lesions (Monakow 1897, 381). Monakow's image thus showed not one single real brain, but was rather a material space of representation in which various scientific techniques had their place, some not carried out on upon the human brain at all.

The development of localisation theory in the twentieth century was no linear success story. Holistic conceptions of an integral function of the entire brain, such as that of Kurt Goldstein, became very attractive for brain research at times, both as a way to resolve certain anomalies posed by the localisation idea and because they avoided the reductionist view of man that, since Gall, had been closely associated with the same. Unaffected by such discussions were further attempts to localise motor and above all sensory functions that were not considered psychological in the narrow sense of the term, yet were of great importance in establishing the iconic presence of mental functions.

The Cortical Homunculus

The interplay of these various considerations can be observed, for example, in the work of Sigmund Freud. In his pre-psychoanalytic study *On Aphasia*, he objected to Wernicke's cerebral localisation of a speech mechanism. He argued that, in spite of the undeniable connection between mental activity and the anatomy of the cerebral cortex, Gall's localisation of the mental faculties and Wernicke's localisation of elementary psychological properties could not escape the arbitrary nature of the formation of psychological concepts. These concepts and categories were based on observations of behaviour and were not necessarily compatible with the complicated laws governing cerebral function (Freud 1992, 96–7, 106–10). Moreover, Freud warned against believing that the individual functions or regions of the periphery of the body had a topographically exact representation in the cerebral cortex. This remark touched upon a very important point: if each individual muscle and each individual area of the skin were projected onto a well-defined location of the cerebral cortex, Freud argued, then eventually a cerebral double of the person herself, however deformed, would inevitably have to appear.

By 1923 Freud had radically changed his view. In a central passage of The Ego and the Id, he argued that the development of the ego is closely linked to the perception of one's own body: "The ego is first and foremost a body-ego; it is not merely a surface entity, but it is itself the projection of a surface". This thesis being untenable without brain anatomy, Freud added: "If we wish to find an anatomical analogy for it we can easily identify it with the 'cortical homunculus' of the anatomists, which stands on its head in the cortex, sticks its heels into the air, faces backwards and, as we know, has its speech-area on the left-hand side" (Freud 1955, 253–4). By this time anatomical, clinical and experimental inscriptions of functions into the cerebral cortex were well underway, yet the little brain man still had the most important part of his career ahead of him.

The first experimental studies were performed exclusively on animals. A whole new field of research was opened up by the first more or less successful neurosurgical operations. Experiments on the human brain were performed in connection with surgical therapy for epilepsy, whereby practical considerations were in the forefront: the electrical stimulation of various areas of the cerebral cortex during the operation were not additional experiments, but rather served the purpose of the exact pinpointing of an epileptic focus. They permitted a more exact mapping, which in turn led to more effective and safe neurosurgical interventions. Thus experimentation and therapeutic treatment were one and the same. Their success was clinically confirmed by post-operative development, and scientifically confirmed by correlations between known anatomical structure and the site of electric stimulation (Hagner 1994). The Breslau neurologist and neurosurgeon Otfried Foerster was able to map a whole series of sensory and motor projection areas onto the cerebral cortex (Foerster 1925).

Results obtained by the direct stimulation of living persons enabled Foerster to enter a whole string of functions into one section of his brain map: "rhythmic chewing, licking, swallowing, smacking movements. Grunting, crowing sounds, hiccups" (ibid., 539–40). The experiments artificially provoked involuntary actions that normally were at least partially under the control of the will. Foerster's patients were awake during the operation, yet they were hardly more than helpless objects of electric stimulation and simulation. Subjected to stimulations of parietal locations of the cerebral cortex, they were expected to experience feelings such as "surging, tickling, electrical sensations, sensations of strangling, muscle fatigue" (ibid., 543). In these experiments carried out by Foerster, the anatomical technique of cytoarchitectonics proved very effective in two respects: first, as a methodological aid to brain surgery, which was now finally showing some promise of success, and secondly, as a re-presentation of the brain's psychological machinery, the full reconstruction of which was the goal of brain research. Above and beyond this, though, the visual images were a manifestation of brain research's claim to truth. For the functional images of the brain had now become more than just a way to mark the location of a researcher's manipulation: they presented truths about the brain and thus about the human being (Fig. 4.6).

It took no more than a few years for a congenial visual substantiation of this claim. The Canadian neurosurgeon Wilder Penfield familiarised himself with the technique of intra-operative stimulation of the cerebral cortex during a research stay with Foerster in Breslau. Penfield had better operational techniques at his command and a far greater number of patients on whom the operations could be carried out. Thus he was able to modify and refine Foerster's picture of the brain, and it turned out that the senso-motoric projections in the cerebral cortex showed the human being to be a being fixated on the thumb, the lips and the tongue. The representation of the human being as a little homunculus (Fig. 4.7) divided into his individual parts marks a new stage in the representation of psychophysical qualities, in which all of a sudden man himself becomes visible, exactly corresponding to Freud's prognosis. The deformations in Penfield's homunculus show that it is now the structural delegation of functions that is considered constitutive of the human being. Prominently drawn body regions like the thumbs, the lips and the tongue indicate that they take up more space in the brain, in proportion to their motor and sensory importance in the world of experience (Penfield and Rasmussen 1950, 114–5). An icon like the homunculus suggests that brain researchers are able to take man apart and put him together again, even if afterwards he looks somewhat different that before. Far from being just a playful drawing, the homunculus is the visualisation of decades of experimental research.

This very brief overview, spanning the period from Gall's organology to Penfield's homunculus, has shown that, while anthropological and psychological concepts and practical procedures have changed, the pretension of being able to understand the mental life of man by quantitative means has not. Gall was concerned with dividing man into properties, desires, tendencies and talents. Later it was a matter of dividing man into psychological and physical functions. The functional brain images used in localisation research steadily evolved during this period, from heuristic orientation aids to a representation of mental qualities or of man himself.

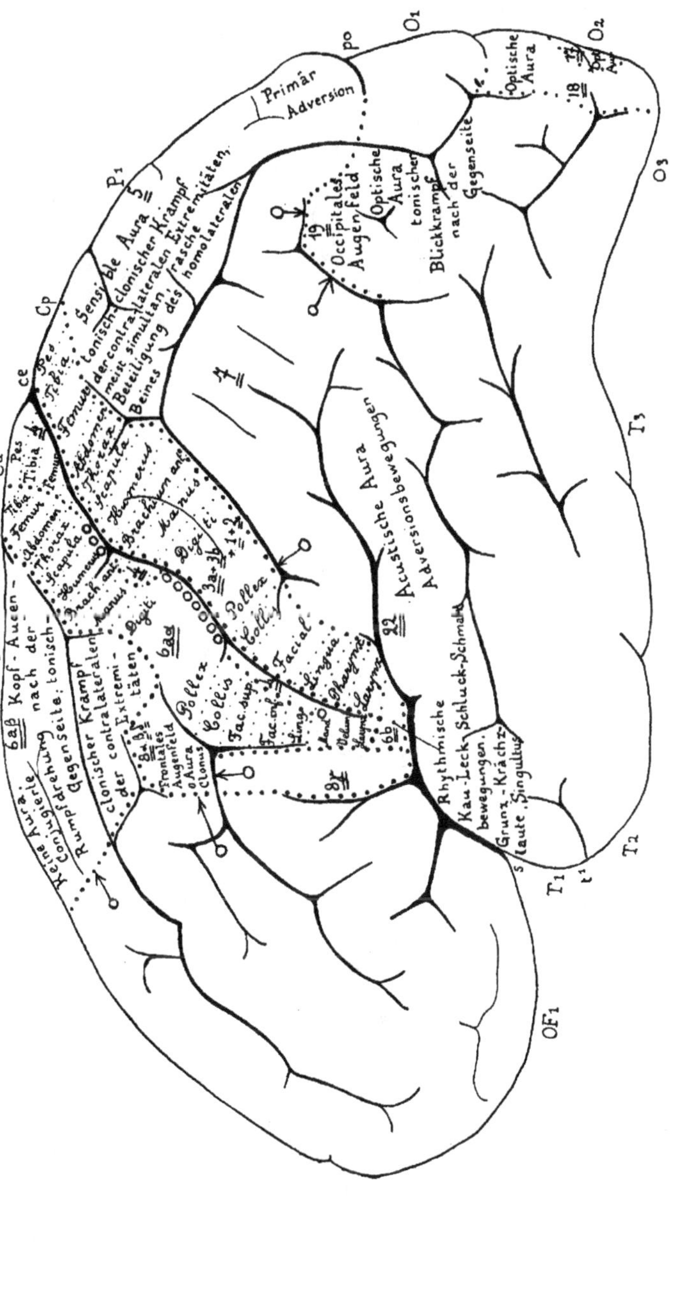

Fig. 4.6. The cerebral representation of sensory and motor centers according to Otfried Foerster. Cécile und Oskar Vogt: Die vergleichend-architektonische und die vergleichend-reizphysiologische Felderung der Grosshirnrinde unter besonderer Berücksichtigung der menschlichen, Die Naturwissenschaften 14, 1926, 1190–1194.

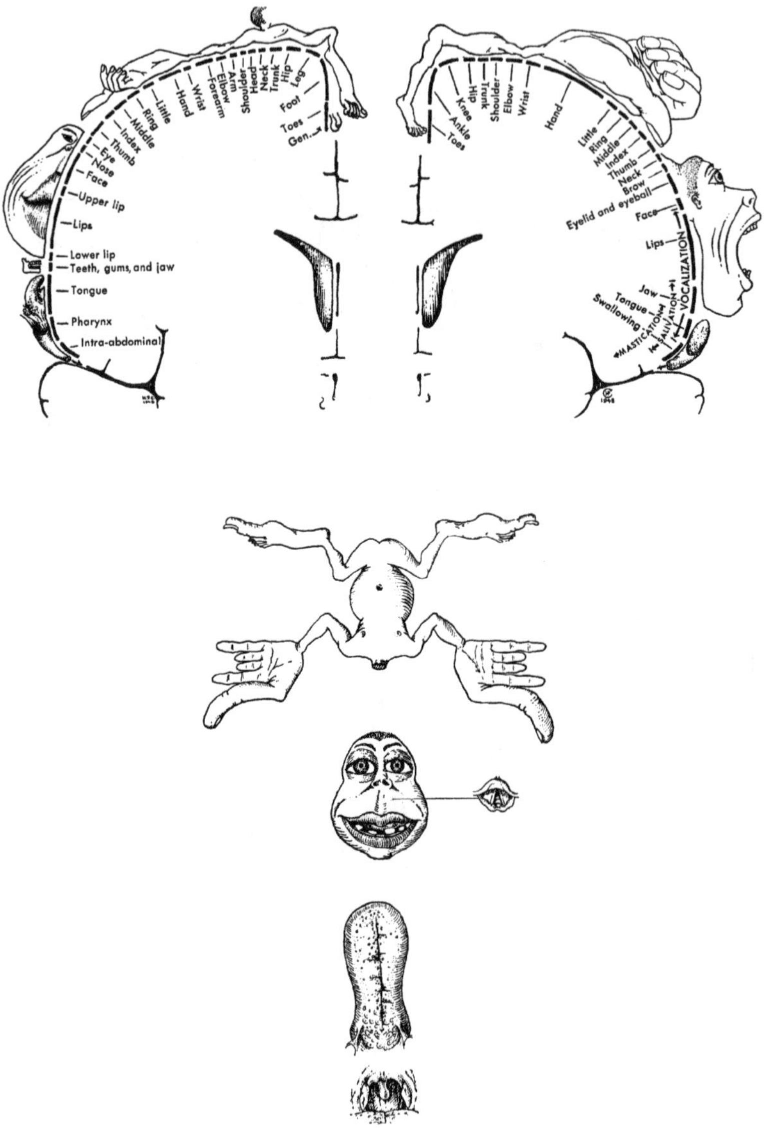

Fig. 4.7. Upper part: Sensory (left) and motor (right) representations of the body on the cerebral cortex; lower part: figurative transformation of the cortical fields into a homunculus (after Wilder Penfield). Wilder Penfield/Edwin Boldrey: Somatic Motor and Sensory Representation in the Cerebral Cortex of Man as Studied by Electrical Stimulation. Brain 60, 1937, 389–443; Wilder Penfield/Theodore Rasmussen: The cerebral cortex of man. A clinical study of localization of function. New York 1950.

Science sometimes does not treat its own history gracefully. The story of scientific progress gets altered to fit current research and its historical classifications. After having been considered firmly anchored textbook knowledge and influencing brain research strategy for several decades, with a flood of graphic representations of the brain in its wake, Penfield's homunculus has been discredited as naïve and misguided. In an article of the influential Journal of Neurology, Neurosurgery and Psychiatry, Penfield was accused of having misused the homunculus to gloss over the methodical uncertainty of his results—for instance, measurements of patient groups that are not comparable, inadequate functional determination of certain areas or failure to differentiate various stimulation types or locations—and therefore of having transgressed the boundary separating fact from fiction (Schott 1993). In other words, the scientific value of the homunculus was cast into doubt, and today it is no longer considered useful or valid.

Some neurologists are a little quicker than the rest of the world. At a time when historians have firmly taken hold of the homunculus as a tangible symbol for brain research's pretension to explain man in his totality, and the little man can be admired in large expositions on the history of research into the human mind, finally acquiring the status of a culturally established "semiophore" (Pomian 1997), the neurologist unceremoniously transplants him back into the scientific past: "For the homunculus, science has now yielded to philosophy" (Schott 1993, 333). This statement can also be understood to mean that the homunculus, no longer needed by science, has now been set loose as an object of free play. It doesn't seem accidental that this releasing is taking place at a time in which the images produced by means of neuroimaging have become the new iconic idols. The homunculus has served his purpose. He can go.

In light of this pointillistic historical account we can pose the question: how do these pictures, sketches and schemata we have considered, including Penfield's homunculus, relate to the computer images of the present day, with respect to their morphological content and their technical production? The answer is simple: they don't. Brain research's pretension to find a scientific solution to the old philosophical mind-body problem has remained constant, and the brain as a material space of representation continues to be reorganized as it has been since the beginning of localization research. But the historically meaningful division between morphological and functional images no longer makes sense today. Whereas former and still current techniques like computer tomography represent structures, and nerve impulses have in the second half of the twentieth century been made visible above all by means of the deflection technique and graphic recording of the electroencephalogram, this division has, with the new techniques of visualisation, become obsolete.

Neuroimaging: Physiognomy turned inward

As the knowledge of brain functions and their localisation is increasingly based on neuroimaging, epistemologically relevant problems arise: what does it mean, for instance, when a certain mental act corresponds to a particular brain image? What does correspond to mean in this case, or the equally often employed term represent? In light of the seductive power of the images: will the symbolic construction of the body and the mind remain acceptable as an answer to the question "What is man"? Or will a new vocabulary be developed to explain our cerebral access to ourselves and the world? How will self-perception change as we humans are increasingly able to voyage within ourselves, not only by means of an endoscope inserted into the natural bodily orifices, but also within our own brains, and thus become able to observe ourselves twice: first by means of introspective thought or feeling, and then by seeing the corresponding cerebral changes on the computer screen? Insights such as these also open up the prospect of a whole new kind of biopolitics. It is almost reality that brain scans will become a means of exteriorising our inner selves in such a way that, in comparison, lie detectors will come to appear like pious father confessors.

Almost 15 years ago, John Taylor, Director of the Centre for Neural Networks at King's College, London, proclaimed: "Windows open on the mind over grey matter" (Taylor 1994, iv–v). That was one of the early examples of the rhetorical rejoicing that has been a determining factor in shaping the public perception of neuroimaging. A central element of such staged productions has been the mutual reinforcement of word and image. In the middle of the double page, next to the title of Taylor's article is a large colour picture that almost looks like a photograph of the Shoemaker-Levy comet colliding with Jupiter: one sees a gleaming fireball on a rather formless sphere that is marked by large furrows and gets darker towards the edges, contrasting with a black background. In actual fact, the picture shows an MRT image. The fireball represents the activated region in the brain of a voluntary subject who is just beginning to silently pronounce a certain word. A brain image such as this, Taylor remarks, brings the mind "into much closer focus". His article "attacks" the ideas of consciousness and self-consciousness like an enemy position, and one learns that the problem of how the brain gives rise to mental processes may soon be definitively solved. Michael Posner and Marcus Raichle, two pioneers of neuro imaging, have expressed similar sentiments in their book Images of Mind, the very title of which holds out the promise of making visible something that is usually considered impossible to visualise. The brain image in the book are already somewhat of a historical artefact, it having become possible in the meantime to represent blood flow much more precisely by means of BOLD-MRT, but that doesn't change the book's pretension to present brave new brain worlds, "the brain systems involved as we think" (Posner and Raichle 1994, 245).

Whereas the phrenologists studied dead fragments of the scull or the brain, the new visualisations of the brain in vivo are a procedure that we ourselves perceive as I think or I feel. All the brain images of the new cognitive neurosciences have a strong physiognomic aspect. One observes the mind at work just as one observes the blushing or the mimicry of a face and draws consequences about the condition of the person concerned. Neuroimaging, then, is a kind of physiognomy turned inward. What do neuroscientists mean when they claim that by accessing neural processes at work in the brain the new imaging techniques are able to produce an ever more refined visualisation of a mental act? Firstly, the old idea of localisation, according to which a particular faculty was associated with a particular area, is now applied to a much more complex mesh. No one doubts the existence of the motor and sensory speech centres. But the questions being asked by present and future research are: which regions of the brain are activated in the course of carrying out a specific linguistic task; which part of the task is carried out in which region; how do the different neuron networks relate to one another? As this research process progresses, it hardly seems utopian to imagine that it will become possible to insert all the thought processes involved, for instance, in calculating, playing chess, composing, making music, practicing a handicraft, reading or looking at pictures into a three dimensional morphological/functional brain image. It has become possible to represent differences in brain activity between the perception of a known and the perception of an unknown face. This possibility may well interest police, the military or secret police. But will it at some time become possible to produce a profile of a person that reads like a CV or an autobiographical sketch, except that the author is no longer the individual reflecting upon him or herself, but rather the visual representation of a certain brain activity determinable in time and space?

Imagine a chess player thinking about his next move. His concentration shows in his posture and facial expression. But these are not specific to playing chess and, of course, say nothing about the content of the thought processes. The brain image showing the mind at work delivers different results. The physiognomy of the brain, the activity pattern accompanying the player's contemplation of his next move, is specific in the sense that it can be distinguished from other mental activities. Furthermore, the activity is quantifiable, at least insofar as for tasks that are the same or comparable, a modest light may flicker for one person while for another a torch glows brightly. Consequences can be drawn from such differences. Even if we assume that data of this kind would not become generally accessible, it is still conceivable that the cerebral profile of an individual could be drawn upon in the selection process for certain sensitive professions and areas of responsibility. At first this would only happen in an additive sense: psychological tests and handwriting tests will not be immediately phased out. But one cannot exclude the possibility that in certain domains brain images could become routine, as has for instance the electrocardiogram in the diagnosis of heart and circulation disorders. That would be above all a question of technical, organisational and financial feasibility. It is true that the electroencephalogram and other methods of examination that have long been in use have not become routine for the determination or optimisation of performance. But that has mainly been due to their comparatively small value as sources of information about the functioning of the brain. By contrast, the electrocardiogram has, in the past decades, played a role that is not to be underestimated in determining suitability for work, vacation, sport or love. Strictly speaking, the ECG delivers information about heart function, but in a wider sense it contributes substantially to the determination of a person's physical performance profile.

For some, the new imaging techniques may be associated with the hope of drawing up a similar mental profile. What would such a hyperphysiogomy bring to light? Disorderly thinking can be distinguished from mathematical problem solving, the recognition of a familiar face from the mute singing of some song, the memories of first childhood experiences from the last fight with one's life companion, the conflict with one's parents from erotic dreams about an arousing love relationship. What is to be won with such differentiations? The answer does neither lie in the brain nor in the machines and is therefore not cerebrally or technically determined. Rather, it can be found solely within the realm of social and cultural negotiations that determine what such information from the brain is supposed to mean.

In the twentieth century probing into the human psyche was first and foremost the task of psychoanalysis. The biographical details, intimacies and hidden layers of the personality unearthed by psychoanalysis will never be equalled by any scanning of the brain. But the crucial point lies elsewhere. Psychoanalysis has had many repercussions, but it never became a standard method for biopolitics. Presumably, this is not because the assumptions of psychoanalysis have been judged to be completely aberrant, nor because the unconscious or human desires are not welcome objects for social engineering. It is much more likely that psychoanalysis seems too elitist and too complicated, too impracticable, tedious and unwieldy. Its results are not primitive enough and too difficult to operationalise for a generalised determination of personality profiles.

Brain images are much more suitable for this purpose. They show much less complex aspects of the personality and are therefore more easily adaptable for biopolitical purposes. Brain images could serve to set norms, consult, control, make suggestions, providing the guidelines for self-appraisals and plans of action. Just as an ECG can give advice on whether a patient is ready for the turbulence of the sex act after having suffered a heart attack, brain images could determine whether someone is more gifted for music or for mathematics. Of course, one image would not suffice for such a purpose. In the course of a certain time period, additional images could be made to determine progress and saved onto a personal data chip, ready to be called up anytime and anywhere. An increase in the lighting up of a student's brain image could act as a motivation for learning, just as an x-ray image shows a patient that his or her broken bone is healing well. The point of reference for the judgement of such images would be the preceding images.

Developments of this kind could lead to a situation in which the diversity and the relevance of mental life are measured too much by its visualisation. According to Vilém Flusser, the new images have "re-enabled us to leave the dissipating world of abstractions behind and return to concrete experiencing, knowing, judging and acting" (Flusser 1992, 44). But there is a price to be paid for this development: "the search for deeper connections, explanations, enumeration, recounting, and calculation, in short historical, scientific, textual linear thinking has been displaced by a new, image-related, 'superficial' way of thinking" (ibid.). In the human sciences, the in-depth probing of the old way of thinking, for which psychoanalysis can be considered representative, is being replaced by the superficial insights of the brain images—without regard for the question of the validity or invalidity of the theses of the former. The study of man would thus have been reduced to the churning out of material forms of representation. It is not a question of the subject having been abolished but rather of the advent of a new anthropology: a cerebral anthropology that in actual fact no longer produces anything but superficial structures, in both senses of the term.

One of the consequences of this new development could be that introspection and subjective interpretations would increasingly fall by the wayside. For some time, various prominent brain researchers have recommended avoiding such categories altogether. Francis Crick calls it an "astonishing hypothesis" that "in order to understand ourselves, we need to understand the behaviour and interaction of nerve cells (Crick 1994, 10). This view is certainly not new, but whereas in earlier times brain researchers retreated at a certain point to positions of scepticism or agnosticism, now joy and suffering, intentionality, identity and free will are taken to be "behaviour of a huge collection of nerve cells and their molecules" (ibid., 17). The mind in its reflective capacity has nothing to say about perception, memory or dreams and the nerve mechanisms on which they are based (Roth 1994, 87). When the world of everyday experience—and for some brain researchers the whole area of the social sciences seems to belong to the same—is severed from a visually produced psychological reality, and when the latter is transported into the world of experience in the form of brain images, then this world will change and with it also the people that live in it.

One cannot prevent anyone from considering the minimal rise in blood flow in a certain region of the brain to be the correlate or even the cause of a mental effort, of the reciting of a poem or of thoughts and feelings accompanying the contemplation of a photographic image of a distorted face. What is to be gained thereby is a question the answer to which is hardly to be found in the images themselves.

Translation: Ursula Froese

CHAPTER FIVE

A PENNY FOR YOUR THOUGHTS: BRAIN-SCANS AND THE MEDIATION OF SUBJECTIVE EMBODIMENT[1]

Renée van de Vall

At the end of an article in the journal *Body & Society* called "How to Talk About the Body?" (Latour 2004), Bruno Latour mentions the neurophilosopher Paul Churchland who carries a colour picture of his wife in his wallet. The picture doesn't show her face, however, but a scan of her brain. In Latour's argument, this example functions to criticise the traditional opposition between the body as an object and as a subject, between the body we have and the body we are, between the known, objectified body of scientific knowledge and the lived, subjective body of phenomenology. Both ways of articulating the body are reductive, according to Latour. Stating that your personality is nothing but the macro structure of your brain would be a very limitative claim. But the opposite claim, that seeing your wife through a brain-scan is a techno-scientific objectification of her personality, is just as reductive because of its denial of the technological mediation of the subjectively lived body. Our usual notions of subjective personhood are also mediated through artificial means and technologies, such as photography. Why not take Paul Churchland's excentricity seriously, adds Latour, and see how he

> is learning to become sensitive, through the mediations of instruments, to hitherto undetectable differences in the spin of electrons of his cherished wife's brain. Paul may be perfectly right in saying that we should all become sensitive to electrical differences in each other's brains and that this sensitivity, this learning to be affected, will make us have a richer and more interesting understanding of other's personality than mere boring facial expressions. (Latour 2004, 225)

[1] I am grateful to the participants of the conference Mediated Bodies and to Robert Zwijnenberg for their comments on earlier drafts of this chapter. Furthermore I thank Simon Cohn for his permission to quote from his manuscript.

The example's rhetorical effect has to do with its simultaneous condensation and denial of at least four oppositions: the opposition between the warmest of feelings, love, and cold sophisticated instruments; between individual personality and general scientific fact; between what is private, accessible only from the inside for ourselves, and public, accessible externally by others; and between what is visible, and therefore evident, and invisible and mysterious because of it. Brain-scans promise to disclose something of our very essence as persons as they open our minds for visual inspection—for ourselves and for others as well. In Paul Churchland's vision of the future—at least as it is implicitly suggested in Latour's account of it—he wouldn't need to pay a penny for his wife's thoughts any longer; they would be as clear to him as if she had told them herself.

The hint remains implicit, but we can see that at least two other oppositions are involved as well, perhaps the most important ones, in particular in their combination: those between mind and body and between subject and object. Brain-scans seem to collapse these oppositions, most spectacularly in those cases when people are able to watch real time functional scans of their own brains, since the subject watches its own mental activity as an objective, bodily process at the very moment that it takes place. In this vein artist Susan Aldworth has told how her fascination with scans of the brain started when she underwent a cerebral angiogram. She was conscious and was able to see the monitors, which, she told later, was one of the strangest experiences of her life. 'Looking up at the screens, I could see the inside of my brain with my eyes—my brain was working, while I was looking inside it. I will never make sense of that moment'.[2] This confusion, this *mise-en-abime*, the vertiginous experience of looking with your brain into your own brain, looking at your own looking, was the starting point of a series of works in which she used brain-scans of various kinds.

I will come back to Aldworth's work and in particular Jenny Slatman's interpretation of it at the end of this chapter. My purpose here is to unpack Latour's example as an instance of what it means to 'learn to be affected', taking it more seriously than Latour might have meant it to be. My question is, what kind of conditions would have to be met in order for Churchland/Latour's vision to become true? Under what circumstances can brain-scans become vehicles for the same kind of feelings as

[2] *The Observer Magazine*, October 1, 2000, quoted in Slatman (2005, 109).

photographic pictures of faces, like for instance being in love with it? It is a hypothetical exercise, but because of its—as yet—virtual character it enables me to outline different dimensions in how techno-scientific discourses and practices mediate embodiment. Cultural analysis tends to assume that in order to understand these mediations it is sufficient to analyse the discursive content of (textual or visual) representations of the body and their institutional embedding, but the question that remains unanswered is whether, to what extent and how those contents actually come to operate in people's understanding of themselves and others. Answering that question requires that one should also investigate how representations acquire personal meaning in daily life.

I will argue that this investigation involves a combination of methodologies that I am tempted to coin 'phenography': an ethnographic-cum-phenomenological description of the material practices in and through which meanings are enacted.[3] A central notion in the phenography of brain-scans will be 'performative articulation', indicating the process in which new (aspects of) worlds emerge, internally differentiate and structure themselves through the embodied doings of those enacting and experiencing these worlds. Although I agree with Latour's criticism of traditional subject/object distinctions, I do not consider each and every use of the distinction erroneous. The terms 'subject' and 'object' may be taken as designating shifting positions that can be occupied by all kinds of actors or actants, human and non-human, as directional poles of experience and agency that can be distributed over a number of entities or internally divide such entities. I will suggest that a closer attention to the ways brain-scans as material artefacts function in people's daily lives and experiences might reveal a more richly differentiated conception of personhood than the semiotic analysis of cultural discourses allows for.

Brain-scans and personality types

That more might be needed than a cultural analysis of brain-scans doesn't mean that we could do without it. A first condition for brain-scans to acquire personal meaning is that we would have to be able to

[3] This term is inspired by Annemarie Mol's notion of 'praxiography' (Mol 2002) and Maud Radstake's notion of 'endography' (Radstake 2007).

read and interpret the images, both the neural activities they denote as their diverse connotations, as part of a shared visual culture. This is increasingly happening; brain-scans of different kinds are present in popular media resulting in a new semiotics of what counts as normal or abnormal. Joseph Dumit has shown how the popular dissemination of PET-scans of brain functions strengthens the notion that there are different kinds of brains belonging to different personality types, such as depressed, schizoid or normal brains, and that these differences can be visualised. (Dumit 2004) This equation of persons with brains, brains with scans and scans with diagnoses is far from self-evident, however. Dumit provides a elaborate account of all the decisions that have to be made in the construction of PET-scans, ranging from the selection of people scanned, the design of tasks to be performed, the choice of radiotracers, to the use of colours to indicate differences in values measured and the selection of images to be published. In medical-scientific contexts, PET-scans are not ascribed diagnostic or evidential authority, but in other settings like popular media, mental-illness communities and courtrooms, they come to have a specific affective and effective power. Outside the medical-scientific discourse they acquire a new kind of evidence, in which the text/image relation is reversed. Whereas in scientific publications, scans function as mere illustrations of an authoritative text, in popular usage the text is reduced to an elaboration of what the images and their immediate labels are supposed to show and prove directly. (Dumit 2004, 142–3)

Dumit's book could be called an ethnographic investigation of the production of cultural meanings. For my question this is not enough. A specific kind of imagery may be prominently present in cultural discourse, on what Don Ihde calls the macroperceptual level of collectively shared (or contested) interpretative frameworks, but absent on the microperceptual level of individual experience. (Ihde 1999, 29) Compare it with the phenomenon of virtual reality, which has been a dominant ingredient in the cultural discourse and cultural theory on new media, but is practically unavailable in day-to-day life. CT, PET of MRI scans may figure prominently in popular medical programs or hospital series, in advertisements or public health campaigns, in court rooms or health policy debates—but this doesn't give them the kind of personal reality that would be necessary for us to 'learn to be affected' by our beloved's brain waves. In order to become part of individual experience, those images have to be affectively incorporated (or rejected) and to be incorporated they have to be encountered on an individual basis.

Part of this incorporation is delineated in what Dumit calls 'the objective-self': 'our taken-for-granted notions, theories, and tendencies regarding human bodies'. (Dumit 2004, 7) Objective-selves are assembled from the received facts of science—that we 'know' that we have brains and that these brains have, for instance, 'reading circuits' might be part of our objective-selves. (Ibid., 7–8) Incomplete and in process as they are, these objective-selves are not articulated for everybody in the same way. The images provided by brain-scans may be vaguely present in people's awareness of what makes up their bodies, but will acquire a different or at least much more pertinent experiential meaning when they are encountered directly, for instance because it is part of one's diagnostic trajectory in a hospital. Dumit provides some very interesting suggestions of how patients might incorporate the kind of 'facts' visualized by brain-scans into their notions of self and personhood—a incorporation that is far more complicated than a simple identification with cultural typologies. People suffering from depression, for instance, might experience it as a relief to identify depression with their brains instead of with themselves as persons and indeed refashion their notions of their personhood in terms of the brain they 'have' as against the brain they 'are', when they picture themselves as persons suffering from a biologically defined illness fighting that illness. The distinction between an actively lived and experienced, subjective body and a passive, medicalized, objectified body does not hold in those cases; what PET-scans reveal is "a medicalized but active, unruly and almost always irrational brain."

> Ironically, the 'normal' brain-type is the one that is, so to speak, passive and lets the real self talk through it. The depressed brain-type, however, substitutes itself for the real self and speaks instead, providing us with such expression as 'That was my illness speaking, not me' [...]. The brain-type, although objective, is simultaneously subjective, lived by the person as well as against the person. (Ibid., 163)

It makes no sense to understand these processes in terms of a strict division of the biological and the personal self:

> Rather the relations between the two selves are redistributed so that although the patient must continue to experience the illness and live with it, she or he no longer has to identify with it. The diseased brain, in this case, becomes a part of a biological body that is experienced phenomenologically but is not the bearer of personhood. Rather, the patient who looks at his or her PET brain scan is an innocent sufferer rationally seeking help. (Ibid., 166)

Dumit's example suggests that patients may construct a very complex picture of who they are, in which simple distinctions between the body as a subject and the body as an object or between lived experience and scientifically induced alienation are, indeed, reductive. Yet these distinctions are not completely redundant. Apparently, patients profit from a kind of 'splitting', through which they separate their 'real' selves, who they are as subjects, from a part of their body, the depressed brains they have. These brains are not merely objects, however, they are both object and subject. As subjects, they are more or less 'other', since they are experienced as part of one's own body but act as separate persons. If this is indeed the case, then rather than the macroperceptual equation of person with brain, brain with scan and scan with diagnosis that prevails on the level of cultural discourse, something more complicated might—at least sometimes—be taking place at the microperceptual level of individual experience.

Body images

The further investigation of this complex identification process could profit from the kind of conceptual framework elaborated by phenomenological studies of the body image or corporeal schema like those of Gail Weiss (Weiss 1999). For Weiss 'the' body is not a pre-given whole, but an active, culturally and technologically mediated achievement. Although the terms 'image' and 'schema' suggest a visual representation, the body image is more like an active principle, a flexible, postural *Gestalt* involved in the integration and stabilisation of bodily perceptions and movements in their interaction with a continuously changing environment. The body image "informs us from moment to moment and in a largely unthematized way, how our body is positioned in space relative to the people, objects and environment around us." (Weiss 1999, 9) But the body image is not a primarily physiological mechanism, as it incorporates psychological and social aspects of the body's orientation in the world. Whereas Maurice Merleau-Ponty has stressed the corporeal anticipation of the body's future actions in and through the body image, psycho-analyst Paul Schilder has emphasized the role of fantasies and the imagination in the construction of the body image. In consequence, there is no such thing as *the* body image, as the way in which people experience their own actual and possible embodied dealings with the world is affected by socially endorsed ideals and negations. Instead, Weiss argues

> [...] for a multiplicity of body images, body images that are co-present in any given individual, and which are themselves constructed through a series of corporeal exchanges that take place both within and outside of specific bodies. (Ibid., 91–2)

Body images are constructed out of a variety of ideals, experiences and notions, including the kind of notions that Dumit has assembled under the concept of 'objective-selves'. These ingredients do not necessarily form wholes, indeed they are sometimes more effective when they maintain a certain multiplicity. In the context of our discussion of the incorporation of brain images, it is interesting that Weiss questions the assumption that fragmentation ('splitting') or (self-) objectivation is always alienating or inevitably brings about oppression. In an appreciative discussion with the work of Iris Young she points to the experience of pregnancy as a resignification of bodily integrity in terms of a splitting that might be paradigmatic for the kind of changes that all bodies undergo. (Ibid., 953) In a comparable manner, patients suffering from depression might restore a sense of integrity by a certain measure of self-alienation, in which a part of their body is separated and recognised as both self and not-self. Objectification and subjectification are played out in an intricate dialectics that is incorporated through the development of a multiple body image.[4]

Learning to be affected

Yet we have not really answered the question how a visual representation succeeds in being *incorporated*, in becoming part of the body's experience of itself and of its orientation and comportment within its world. Dumit's research does not include ethnographic observations of how patients actually achieve their identification with scans, and although Weiss stresses the performative dimension of the development of body images and their function in the body's 'becoming', she doesn't provide

[4] Maud Radstake has suggested replacing the concept of body image by Ihde's concept 'image-body'. (Radstake 2007, 116–117) I would rather see Ihde's distinction of 'here body' and 'image body' as an internal differentiation of 'the' body image. What Mol has stated about 'the' body being a multiple (Mol 2002), goes for the body image as well. Body parts that are experienced as self or partly self and partly not-self, might likewise be perceived as from within (here body) or as from without (image body). (Ihde 2002, 4–5) Mark Hansen has convincingly argued for a sharp distinction between the concepts of body schema and body image. (Hansen 2006) Although I agree with his argument, applying the distinction in this chapter would not alter its conclusions.

a framework for the analysis of this dimension. Here I think lies the contribution of the ethnographical, Actor-Network-Theory approach to cultural theory and phenomenology. When we would limit ourselves to the critical analysis of the politics of representation, as has long been the dominant practice in culture studies, we would miss the differentiated ways in which images are actually functioning in various practices. Of course, cultural studies have also foregrounded decoding practices and what they do to confirm or counteract dominant ideologies, but ethnography goes further than that in its empirical investigation of what Annemarie Mol calls the 'practicalities' of these practices. (Mol 2002, 31) These practicalities may differ substantially and with that, what images come to mean.

For Latour, "to have a body is *to learn to be affected*, meaning 'effectuated', moved, put into motion by other entities, humans and nonhumans." (Latour 2004, 205) It makes no sense to describe the body in isolation from the practices and technologies in which it partakes; nor to distinguish a subjectively lived body from an objectified, instrumentalised, scientifically 'known' body. Learning to be affected operates through the progressive articulation of what Latour calls 'propositions'[5] and is mediated by all kinds of people, tools, and practices—hospitals and scanning technologies not excluded.

Latour's example of how a body learns to be affected is the training of 'noses' through a so-called 'odour kit' by the perfume industry. The odour kit consists of series of contrasting fragrances, which one learns to register in a week-long session. People who at first could only distinguish 'sweet' from 'fetid' odours become able to discriminate more and more subtle differences and to tell them apart from one another. They become 'noses'. What was an inarticulate experience—the registration of sweet or fetid odors—becomes an articulate one—let's say the registration of a combination of violet and musk—that is, an experience that, instead of creating a "general undifferentiated effect or affect", makes those who have learned to be attentive to "ever subtler differences in the inner structure" of what is experienced, "do something different every time—instead of eliciting always the same crude

[5] In contrast to a statement, epistemologically defined as 'a sentence that is judged to be true or false', a proposition is ontologically defined by Latour as 'what an actor offers to other actors'. (Latour 1999, 309)

behaviour." (Ibid., 205) In other words the crux of the articulation is in the *doing*: the activities that make a new articulation possible and those it subsequently solicits. In the process, both bodies and worlds change—learning to be affected is *becoming* a nose and progressively inhabiting a more interesting and complex world.

What goes for the perfume industry goes for science as well. The paradoxical result of much science studies concerned with the body, Latour concludes, is neither a fight against reductionism nor a plea for the whole personal, subjective body that should be respected instead of cut into pieces. Rather it is to show how impossible it is to be really reductionist, even for the most reductionist scientist and even in the most objectifying setting, like a hospital. Hospitals feed articulations. Instead of being reduced to "a mere pack of objective meat", you are

> learning to be affected by masses of agencies hitherto unknown not only to you, but also to doctors, nurses, administrators, biologists, researchers who add to your poor inarticulate body complete sets of new instruments—including maybe CAT-scans. (Latour 2004, 226–7)

Learning to be affected through medical technologies

One can learn to be affected by one's own body as well. In the same issue of *Body & Society* Annemarie Mol and John Law tell how hypoglycaemia, a low blood glucose level which may occur to people suffering from diabetes, can be measured by taking a blood sample and putting the blood into a machine, but can also be felt—people who are sensitive to their own physical state can feel it coming on. Self-awareness, the chiasmatic feeling of oneself feeling, is important in the day to day handling or avoiding of hypoglycaemia, but, Mol and Law stress, it is not straightforward. It is stronger in some people than in others, in some it is lacking, in others it is inaccurate. Moreover it can be *trained*—with the help of measurement machines. People train their inner sensitivity in training programmes in which they first guess their blood sugar level and then measure them. By starting from the body we *do*, Mol and Law are able to show a very interesting relation between the subjectivity of the body we are and the body we know objectively: the former is used as a resource and even extended by the latter. (Mol and Law 2004, 47–8)

Like measurement machines. visual images of the interior body mediate articulations of the body for the body. Maud Radstake has investigated how patients encounter real-time medical images[6] of their insides in a hospital and how they do or do not connect these images with themselves—their bodies. There are three stages in the encounter: the preparation for the imaging event, the event itself and the diagnostic trajectory after the imaging. Before the visualisation procedure can take place, patients have to actively prepare their bodies, for instance by drinking water (in case of an ultrasound of the lower abdomen), or by refraining from eating and taking laxatives (in the case of a fluoroscopy or endoscopy of the intestines). (Radstake 2007, 54–60) In the imaging event itself, Radstake distinguishes four different, but interlinked modes of attachment: the real-time interpretation of images by medical professionals; patients' visual perceptions of their own interior body and their simultaneous sensory perceptions of the imaging procedure; patient's physical agency during imaging; and patients' emotional responses to the images and the imaging procedure. (Ibid., 80) These modes may differ from technique to technique, but also from encounter to encounter, Radstake noticed, depending on the actual behaviour of doctors and patients. With regard to the real-time interpretation of images by medical professionals she observed:

> Sometimes, doctors actively involve patients in the productions of images and in watching them. Some physicians explicitly point to the possibility of watching the monitor during an examination. Yet during the examination, they may not explain anything about what is shown. Others give almost a streaming commentary. (Ibid., 85).

Some patients actively observe the images and discuss them with the doctors, others refrain from looking at them. Patients who do look may not recognise what they see, but as they also feel the imaging procedure, for instance because they feel the endoscope moving inside, they connect what they see with their own body. They also may react emotionally, feeling embarrassment or fear, but also awe or fascination, sadness or loss. (Ibid., 87, 91–2).

After the imaging event, the real-time connection of the image and the body is broken. Most of the images produced disappear after the examination. The events are translated into reports by the medical professionals and into narratives by the patients. Both translations

---

[6] The technologies Radstake has investigated are ultrasound, fluoroscopy and endoscopy of the abdomen.

cross ways and shape each other's direction. Radstake describes how real-time visual images become part of what Dumit has called patients' 'objective selves':

> In the meantime [before the follow-up visit to the attending physician, RvdV], doctors have discussed their findings, edited and stored stills, dictated reports and checked the written version. Assistants have sent tissue samples to the laboratory where they have been translated in more numbers and words. [...] patients take their memories of the examination home with them: the sensations, the pictures, the doctor's explanations. Awaiting the diagnosis, patients translate their experiences during real-time imaging events into mental images and into words and stories told to relatives, friends or an inquisitive anthropologist. Patients' experiences include their observations of professionals performing the examination. The result of the examination they eventually receive through their attending physician some days or weeks after the event, becomes part of patients' experience as well. (Ibid., 102)

'Learning to be affected' by images of one's own insides, whether one's intestines or one's brains, may take various forms, as Radstake shows, depending on different variables, such as the material equipment used, the spatial arrangement of the technologies involved and the way both patients and doctors respond to the equipment's possibilities. Her investigations and reflections point to another practicality with experiential consequences. The visualisation techniques in her research produce real-time images that exist only temporarily on screens, and seldom as the kind of prints husbands may carry around in their wallets. After the imaging event, the images exist only in translated forms for the patients, as mental images or as words and stories. The same goes for brain-scans: they seldom leave the hospital—unlike foetal ultrasounds, for instance, that are taken home and shown to family and friends.

But sometimes they do. Anthropologist Simon Cohn has interviewed patient-volunteers who, after having participated in scanning experiments were given printouts or CDs with their brain-scans. His findings provide a fascinating insight in the dynamics of the performative articulation of illness and the importance of the materiality of mediating actants in the configuration of patients' experience. His account partly confirms Dumit's suggestions. For many of the patients, the scans were given an active role in the re-conceptualisation of their illness as having a physical basis, which could be localised inside the brain, yet could be separated from the self or the person they were. Not only did they consider the printout as they would any photograph, as a direct visual testimony of what happens inside. The scans were also important as material entities one could do things with, especially in the social sphere:

> For them, the technology allows a re-articulation of what is conceived of as internal, and thus only personal, with what can be made external and consequently an object with which to shape relationships with others. (Cohn forthcoming, 17)

Cohn mentions a woman who carried a copy of the scan in her purse, telling him that she found it 'really comforting to have it with me', others put scans into albums, frame them and put them on the wall, or print them on T-shirts. He comments that it is striking 'how common the desire to actively do something with the image is, as though making the illness concrete and distinct at a conceptual level is not enough.' (Ibid., 19)

> It is as though the many various activities serve as both means to appropriate them and convert them, so that they can be incorporated into some part of people's life as something that is active and performative—whether that is simply a picture on the wall, a tee shirt or even a place-mat. [...] The work that many patients do to the scans can be interpreted as a method to apply their own personal agency to them, such that the appropriated image acts as an extension of their themselves in the social life that then surrounds them. In this way, the scans become dynamic objects that can be used to initiate new ways not only for themselves to think about their own condition, but also potentially serving as invitations for new ways other people can be engaged too.' (Ibid., 19)

We could say that this is an example of the way not only people, but also things can 'do', shaping the ways in which people relate to themselves and others. (cf. Verbeek 2005) Without the scans patients would not be able to experience their illness in the particular way sketched above, nor to communicate in this way about it. However, it is not the scan in itself that has this agency; it is the scan as involved in a practice. And what it does is not always the same. Cohn also encountered patients who, after a while, became disappointed because their social world did not respond to the scans as they had hoped. Then the semiotic connotations of the brain-scan may actually become a trap, because of the scan's supposed indexical value as a proof of their condition's physiological basis. In other words, brain-scans may feed articulations, but also freeze them into stereotypes.

Art as performative articulation

> I still cannot escape the thought that I am seeing all this and thinking all this because of the very thing that I am looking at. The brain is a very strange and marvellous thing. It is not like a heart or a kidney. It is thinking flesh. No wonder I keep changing materials for my work. I don't have a problem with its structure—it is function and consciousness which baffles me.
>
> Aldworth's diary describing her experience of a cerebral angiogram, posted on October 26, 2001. (http://www.susanaldworth.com)

One of the most difficult theoretical problems connected with the empirical approach of ANT ethnography is that it has to address normative questions without relying on transcendent values, like the 'truth' of representations or the 'wholeness' of embodied experience. When some patients in Simon Cohn's research enact specific identities in their dealings with scans that finally only entrap them, the reference to the truth or untruth of the scans is not a theoretical option consistent with the ANT approach. There seem to be no independent standards enabling the criticism of practices. As Annemarie Mol has remarked, in an empirical approach the 'other' to a practice is always another practice.[7] However, sometimes such alternative practices exist. I would like to point to art as a particularly informative type of alternative and potentially critical practice. Both as productive and as receptive practices artworks embody forms of performative articulation. As such, they possess the same kind of reflective potential as the ethnographic and phenomenological studies this article has drawn upon, but in contrast to those studies, this reflective potential is exemplified in the same kind of visual/material form as the artefacts on which it reflects.

According to Jenny Slatman, the central problems addressed in Susan Aldworth's works[8] are best captured by the title of one of her expositions, *Between a Thing and a Thought.* (Slatman 2005a) Are brains matter or mind or something in between? Out of that question more questions follow: questions of identity—who I am for myself and

[7] In a abstract for the conference *Mediated Bodies*, Maastricht University September 14–16 2006.

[8] Susan Aldworth works in etching, collagraph, digital print and film. Her current practice is interdisciplinary and inspired by medical science and working on location in hospitals. Her most recent work focuses on the human brain and issues of personal identity. (www.saatchi-gallery.co.uk/yourgallery/artist_profile/Susan+Aldworth/11773.html)

others?—and about temporality—to what extent can a (functional) brain-scan visualize 'me' or 'you' by highlighting a single moment of mental activity? On the one hand we 'are' our brains and in this respect brain-scans make 'us' publicly visible for ourselves and others (actually, one of Aldworth's works is titled *A Penny for Your Thoughts*); on the other hand we are many other things as well. In each of the three works of the series *Cogito Ergo Sum* Aldworth has used twenty small MRI-scans arranged as in a cartoon. Read from left to right and top to bottom, the scans are increasingly elaborated with added drawings figuring eyes, faces, figures or dark spots, and with texts like 'You can look INTO my brains but you will never find me' or 'I am both in my head, and out of my brain'.

By articulating the confusions that arise from the possibility of visualizing mental activity, and by inserting the visualizations in a sort of narrative, Aldworth reconnects medical images with daily experience and the flow of time, according to Slatman. With that, we could say that her work is an example of someone who, like Paul Churchland and Cohn's patients, but with different means, is 'learning to be affected'. But rather than articulating the one-to-one relationship between scan and brain and brain and person that the Churchland/Latour example suggests and some of the patients in Cohn's research finally were entrapped by, Aldworth's reworkings of the scans question such equations. They do so not by evoking transcendent values or standards, but by presenting the scans in unusual ways and connecting them with other images and imaging techniques. Here the multiplicity of bodies and images, subjects and objects, interiority and exteriority is explored in an artistic way; the tensions between different enactments of being me and relating to a you through a brain scan are negotiated and conveyed in a visible form.

Neither in real life, nor in this article we have reached the point where we might fall in love with a brain-scan. What I hope to have shown, however, is how a new imaging technology like brain-scans may come to mediate bodies. I have argued that this requires that those scans function on the microperceptual level of individual experience and the more so if and when they leave the hospital and are (literally) transported into daily life. To the extent that brain-scans figure as material artifacts in the performative articulation and social enactment of subjective embodiment, they may even complexify and enrich our notions of our own and others' personalities—which would not be apparent if we would only look at their more general cultural significations. Whether they actually do so or not is not given in a once and for all way. The drawback of the theoretical and methodological approach chosen—the combination of phenomenology with ANT-ethnography—is that it does not allow for independent standards of judgment formulated on a meta-level. However, alternative practices involving brain-scans, such as artworks, may be explored to discover the potentials and limitations of specific uses of imagery.

CHAPTER SIX

TRANSPARENT BODIES: REVEALING THE MYTH OF INTERIORITY

Jenny Slatman

Introduction

Since the invention of x-rays in 1895 by Wilhelm Roentgen, the interior body has become accessible and visible without dissection. At the turn of the previous century, Roentgen's invention marks a new area, which I would call the Enlightenment of the body. The body can be entirely screened and elucidated by shining light through it. Especially since the 1960s, the development of all kinds of imaging technologies, such as endoscopy, ultrasound, MRI, PET and CT scans, has expanded enormously. It is characteristic for these technologies that they can fairly easy enter the body by letting its surface intact and undamaged.[1] Although current practices of illuminating the body can be seen as a logical continuation of the will to knowledge, having its origins in the eighteenth century of Enlightenment, it is a crucial difference that we can now make visible the interior body while it is still alive. Moreover, we can even view the interior of our *own* living body.

Of course, these technologies are primarily developed to increase medical possibilities of examination and treatment. But they also change the way in which we perceive and think about the body. It is a widespread conviction—in Western culture—that this kind of imaging renders the body totally comprehensible, or at least, it triggers the belief of a transparent body (Van Dijck 2005). It would be silly to deny that contemporary imaging unveils numerous anatomical 'secrets', which were formerly hidden. I would like to show here, however, that it is

[1] Needless to say, this does not mean that these interventions are always painless. Some of them can be rather exhausting and agonizing. It is not very pleasant to get an endoscopic camera in the intestine or urethra. Moreover, some forms of imaging leave cicatrices, as is the case by some applications of endoscopy, when a (small) incision in the skin is required.

inevitable that these illuminating practices face a specific kind of opacity in their search for transparency. My claim is that although these imaging technologies might travel within living bodies—our own living bodies—they cannot truly represent a living body. They do represent the body's interior space, but as I will make clear, this unveiled inside is not the hiding place of something like the living body. Since medical imaging always encounters some opacity, the idea of a complete transparent body is just an illusion. I will add to this that it is because of this opacity that we even have to give up our current idea of interiority. We are so used to speak of our 'interior', and it might seem that modern imaging technologies make this easier, but, in the end, they only affirm that such an idea is nothing but a myth.[2]

At first glance, we might be inclined to say that interiority is the space beneath the skin. The skin is the demarcation between the inside and the outside. Naturally, we can speak of the part of our body that is beneath the skin as something interior. That the body is a kind of container with an interior is clear by the very fact of the existence of the orifices such as the mouth and the anus: we can put something in our body and we can excrete things. In the same way, the endoscopic camera literally enters the body: it literally enters the space beneath the skin. The possibility of entering the body with a camera has triggered our imagination; our own body can be thought of as a place that can be entered and in which one can travel.[3] And yet, I do not think that if we speak of the body's interiority we only allude to a space that is demarcated by the skin. The body's interior is not just the same as the interior of a house demarcated by walls. Somewhere deep 'in' my body—perhaps in my brain, or perhaps not really localisable, or perhaps

[2] The term 'myth' is borrowed from Gilbert Ryle (1949), who argues that Descartes' idea of the existence of the mind as an independent substance, opposed to the physical substance, is nothing but a myth. Ryle's disclosure of Descartes' myth implies a 'demystification of the mind': the mind should not be understood as an enigmatic inner substance independent of the body; the mind is not a 'ghost in the machine' (p. 17). In this paper, I would like to contribute to this 'demystification', although my approach cannot directly be compared with Ryle's. His demystification of the mind proceeds from a conceptual analysis of our ordinary mental vocabulary. According to him, the myth is based on a category mistake. I will disclose the myth by means of a phenomenological analysis of touch and vision.

[3] The movie *Fantastic Voyage* (1966) is a very nice example of this imagination. Nowadays, the use of pictures and animations of the inner's bodies (travelling) space is an approved method in commercials that recommend all kind of self-help medication or healthy nutrition.

only symbolized by my heart—there is something that we usually call our inner self. This inner self refers to an interiority that is not spatial. I would say that it is psychological. In my opinion, if we want to explore the meaning of the *living* body we cannot limit our selves to the spatial meaning of interiority, but should as well take into consideration its psychological meaning.

In general, psychological interiority stands for our inner feelings, thoughts or states of mind. Interiority in this sense refers to that part of us that is considered to be our most private part and even the part that constitutes our very unique being. Whereas my outer appearance, my overt behaviour is visible and observable for every one, the access to my interiority seems to be restricted to myself. If I have certain thoughts, no one knows what I am thinking. If I have pain, no one knows exactly how my pain feels. If I enjoy my coffee, no one knows exactly how I enjoy it. Or at least, this is how we normally conceive of our inner states of mind. I do not think that contemporary imaging technologies, however invasive they are, really penetrate the interiority of inner feelings and states of mind. Even if inner states of mind can be reduced to brain activity, as some neuroscientists and philosophers claim,[4] and brain activity is registered by a scan, it does not follow logically from this that inner states of mind are made visible. Undoubtedly, inner states of mind are related to brain activity, but still, if one registers my brain activity, I do not think that this will make visible how my pain feels or how much I enjoy my coffee.[5] Medical imaging is thus not applied for the exploration of a psychological interiority. Rather, it is used for the examination of spatial interiority. A physician might be interested in, for example, whether there is a polyp in the intestine, whether a foetus is growing sufficiently in the mother's womb, or whether sudden speech failure is due to a vascular accident in the brains.

It is thus clear that the psychological meaning of interiority does not match immediately with its physical, spatial meaning. And yet, in

---

[4] See for instance Churchland (1986).

[5] According to the vocabulary of contemporary *Philosophy of Mind*, one would say that it is not possible to approach *qualia* from a scientific, third person's perspective. *Qalia* refer to the quality of experience; they refer to what it is like to have mental states. Theorists claiming that inner states of mind should be reduced to brain activity, normally tend to deny the existence of *qualia*. Although it is not my intention to get entangled in the *qualia*-debate here, implicitly I will argue that we cannot deny something like the quality of our experience, but this quality is not something subjective or private, not something that stems from some kind of interiority.

this paper, I will show that psychological interiority must be related to some kind of spatiality. We can have no feelings, no experience at all without a spatial body. Psychological and spatial interiority converge in the so-called phenomenon of bodily subjectivity. It is only on the basis of this phenomenon that we can understand what constitutes a living body. I will explain it as the bodily experience or sensing of one-self. Although I consider the body in terms of experience my approach is not psychological but rather philosophical. While drawing on some ideas of Descartes, Husserl, Merleau-Ponty and Nancy, I will provide a twofold argument. Firstly, I will make clear that 'psychological' interiority, i.e. our subjective feelings and thoughts, is not just a mental thing, but something that is inherently corporal. Secondly, if interiority or subjectivity can be considered as something corporal, it can no longer be understood as something that is just my private 'place' or something that I own. Something strange and exterior persists, even in our inmost inner feelings.

On the basis of this idea of embodiment and interiority, I will explain that there always escapes something from the picture in contemporary imaging technologies. However transparent these pictures can make our inner body, and however lively these pictures might look to us, they cannot grasp the living body. A picture is made of our inner body while we are alive, and yet, it cannot show anything else than our thinghood, which in essence is the inanimate aspect of our being. It is the inanimate that resists complete transparency. Picturing living bodies or 'dissecting' living bodies by means of an invasive (harmless) gaze remains a science or a practice of dead bodies, or at least, of the dead, inanimate side of the living body.

Interiority and Subjectivity

It is not surprising that interiority has not only a spatial meaning. Taking into consideration the etymology of the term, we see that it is related to intimacy. In Latin, *intimus* is the superlative degree of *interior*. Intimacy is profound interiority. Someone's intimacy lies in someone's inmost interiority. As said before, it is not a public but rather a private side of me. Or to phrase it differently, something intimate is supposed to be private and not public. Whenever something intimate is made public, its intimacy is violated. The intimate belongs to me, to my personal sphere. It is therefore that we can say that it is our inmost interiority, our intimacy that constitutes my personal and proper being. Interiority in this sense indicates my self, my most genuine me-ness. It is not by accident than that the very idea of a self or a person has always been related to the assumption of interiority. In western philosophy, the very idea of personhood was only developed after Saint Augustine's *Confessions*, in which he scrutinizes his inner self by descending in his memory.

In modern terms, interiority stands for subjectivity. Although various contemporary philosophers underline that subjectivity should be distinguished from what is called 'substance', the term subject is derived from the Latin *substantia*, a translation of the Greek *hupokeimenon*, which means 'that what sustains, what remains the same'.[6] Subjectivity, therefore, refers not only to the personal or the individual, but also to something—be it a substance or not—that remains stable and the same. If my interiority stands for my subjectivity, it stands for the part of my existence that remains constant—in contrast with my ever-changing looks (at the outside).

Probably Descartes provides the clearest example of the coincidence of subjectivity and interiority. It is often said that he is the 'father' of modern (western) thought, since he 'invented' the idea of a subject. We must note here immediately that Descartes has never used the term 'subject' himself—he sticks to the then prevailing term 'substance' (*res, chose*), which he applies to both physical (and mortal) beings and the being of the immortal mind (as well as to the specific being of God). However, what he called the mind (*esprit*) or the thinking substance (*res cogitans*) is exactly what later came to be known as the subject. The 'birth' of this subject in the seventeenth century does not only mark a crucial shift in philosophy. It was equally important for developments in psychology. It is not exaggerated to state that our modern and contemporary idea of mind or *psyche* can be related immediately to Descartes' conception of the *res cogitans* (Fancher). Also methodological quarrels between behaviouristic approaches and approaches based on introspection have their origin in Descartes' philosophy. The very idea that my mind is something private stems from the Cartesian idea that the true sense of the mind can only be discovered by means of meditation, that is by means of a look 'inwards'.

[6] In my discussion of Nancy's conception of the body, here below, I will return to the distinction between subject and substance.

In his *Meditations*, Descartes invites us to get ahead of our usual way of thinking. To find an indubitable basis for all sciences, we must put into brackets everything that allows the smallest doubt. Since perception and imagination can provide doubtful information, we should not trust these faculties. Even rational knowledge of which we are pretty sure, such as the fact that 2+5=7, can be put into question, since we never know whether there is an evil genius that misleads us all the time and that makes us believe that 2+5=7. Indeed, everything that we consider to be true in our daily life can be doubted. Everything can be doubted, except for one thing; the very fact that I am thinking: *cogito*. Doubting is a form of thinking, and therefore, if I doubt, I am thinking.[7] At the very moment one puts into doubt everything, one cannot have any doubt about the existence of one's *ego* that performs the act of thinking. Even if everything can be doubted, there must be someone who doubts. Consequently, Descartes argues that it is the 'I' (*ego*) that thinks that constitutes an indubitable base. I will leave aside his arguments about further foundation of knowledge and only concentrate on the way in which Descartes describes this indubitable ego. According to him, this *ego* has no physical properties, since everything physical can be doubted. For the same reason, it cannot even be imagined (for, everything that can be imagined can be doubted). The only 'property' that can be ascribed to this *ego* is the actual pursuit of thinking. Only the act of thinking itself—and not the object of thought—is indubitable. As I see it, there are two things that are significant in Descartes' way of reasoning. Firstly, he proposes an absolute freedom of thinking. Thinking has no constraints: everything can be doubted. Secondly, this unrestricted way of thinking implies a withdrawal from everything exterior to me (since everything outside me can be doubted); a renunciation of the world. Indeed, to meditate means to close one's eyes for the exterior world and to turn one's mind to one's inner self. My indubitable *res cogitans* can only be experienced by myself by means of introspection and cannot be observed from the outside. This leads to the conclusion that my *cogito* is something interior that is enclosed upon itself.

Descartes' quest for something indubitable has led him into man's interior, which goes together with a complete abandonment from the body; for, anything physical can be doubted. As we know, this kind of reasoning has resulted in a strict dualism between body and mind,

[7] "But what is a thinking thing? It is a thing that doubts, understands, [conceives], affirms, denies, wills, refuses, that imagines also, and perceives" (Descartes 1988, 34).

between the exterior and the interior, which, of course, has raised many questions about how the two can be related to each other. This problem, albeit rather fundamental in modern philosophy and psychology, is not of my interest here. What interests me here is that Descartes' description of subjectivity in terms of an indubitable *res cogitans* implies an interiority that is opposed to anything physical. Interiority in its psychological sense, so it seems, has nothing to do with the body, with a corporal space. This Cartesian way of thinking—which is still predominant in our culture and science—is based on a very limited and simple idea of the body, that is to say the body as a pure thing, extended in space (*res extensa*). Since this body is no more than a thing or an object, it is the opposite of the subject. Contrary to this (predominant) view, I will show that the body can be understood in a more complex way. If we concentrate on the experience of our *own* body, we not only experience it as a thing, that can be replaced by any other thing, but also as something that is very intimate and personal. For instance, when I feel pain, this does not mean that my body causes pain (in the mind), but it means that *my* body hurts. Descartes already pinpointed at this close-fitting relation between the body and the experience of ourselves when he stated, with a reference to Aristotle, that the relation between the mind (that experiences) and the body should not be thought of in terms of a pilot that lodges in his vessel.[8] Still, Descartes has never explained this bodily experience of our selves in a convincing way. The reason for this is that he remained attached to the idea of the body as a thing. Various 20th century philosophers have criticized this idea of the body. In the next section, I will explain in which way the body can be thought of in terms of subjectivity instead of in terms of thinghood or objectivity.

Bodily Subjectivity

To my knowledge, the first philosopher who speaks of a bodily subject is Merleau-Ponty in his *Phenomenology of Perception* (1945). However, for his description of the *corps sujet*, he cites, among other things, the

[8] "Nature likewise teaches me by these sensations of pain, hunger, thirst, etc., that I am not only lodged in my body as a pilot in a vessel, but that I am besides so intimately conjoined, and as it were intermixed with it, that my mind and body compose a certain unity" (Descartes 1988, 94) See also Aristotle's *On the Soul* (413A).

analysis of touch provided by the German philosopher Husserl. To understand what can be meant by bodily subjectivity I will therefore first explore this analysis of touch. An additional reason for starting with a reading of this German text is that the German vocabulary provides a very adequate distinction with respect to the body—the distinction between *Leib* and *Körper*—which is lacking in French (and English).

In a passage of *Ideas II*, Husserl describes what happens when two hands of the same body are touching each other.[9] When my right hand touches my left hand, the right hand can be considered the active touching hand. The left hand, by contrast, is the passive touched hand. On the level of experience we can distinguish here two aspects; let us say a subject and an object side. The experience provided by the active hand forms a subject. This is not yet what I would call a bodily subject since the information that stems from this touch experience, such as information on physical qualities, could also be considered as mental, cognitive experience. So far, nothing new. The touched hand is experienced as an object. This also does not break with traditional views. The innovation of Husserl's analysis lies in the fact that he recognizes that the touched hand is not just the same as any other touched object. There is a crucial difference between my right hand touching, for example, a keyboard and my right hand touching my left hand. My touched hand can feel that it is touched; it can feel its own touchability. The touched hand feels its touchability as localized in this touched hand. The feeling of one's own touchability marks the shift from the body as a *Körper* to the body as a *Leib*. A *Körper* is a thing with physical qualities. A *Leib*, on the contrary, is not a thing extended in space with physical qualities. Although the *Leib* has a certain spatiality, since it is the localization of touch-sensations, these localized sensations—which are named 'sensings' (*Empfindnisse*)—do not constitute physical features such as smoothness or roughness and they thus do not constitute the physical thing 'left hand'. The 'sensing' does not provide the state (*Zustand*) of the material thing hand; it is thus not the perception of my hand as a cold, smooth or rough object. Rather it provides the hand *itself*, the hand as *my* hand, which "for us is more than a material thing" (Husserl 1989, 157).

The body as *Leib* is constituted by auto-affection; the experience of sensing oneself as being sensed. Sensing oneself as being sensed

[9] See especially § 36 of the *Ideas II*.

refers to another dimension than sensing oneself as a certain object. One might be inclined to state that this feeling of oneself is nothing other than a mental feeling of oneself and in that sense nothing else than Descartes' *res cogitans* feeling, for instance, pain. Then, the *Leib* is perhaps nothing other than the mind. In a sense this is true, but there is a crucial difference. If the *Leib* were a pure *Leib*, it would be nothing other than a mind or a *res cogitans*. However, it is essential to the *Leib*, being the experience of one's own body, that it is inextricably linked up with the *Körper*. Since *Leib* stands for the sensing of one's own being sensed or touched, it cannot exist without the very possibility of touchability which is provided by the *Körper*. In that sense, the *Leib* simultaneously presupposes and affirms our *körperlich* being. Without a *Körper* there is nothing to be touched and thus nothing that can be sensed as being touched. In Cartesian terminology this means that the *Leib* might be compared with a mental experience of sensing, but only to the extend that this experience cannot be stripped off its physical *res extensa*. In the experience of my own body, provided by the experience of touch, something like the *Körper* cannot be eliminated, even though it seems to be doubtable. We should perhaps argue against Descartes that although we can doubt about nearly every outer appearance of our body, we cannot doubt about the fact that this very body that is touched is *my* body, and to have such an experience of me-ness, of subjectivity, we need to have a touchable body. The body can thus be seen as a subject (in the sense of *Leib*) but a subject that is never pure. This idea of impure subjectivity has important consequences for our further analysis of the interior body.

The fact that the subject or the *cogito* can never be purified completely from its physicality has, at least, two important consequences. Firstly, this subject does not coincide with itself, since the experience of being touched can only occur if there remains a difference between *Leib* and *Körper*. If the two could coincide there would be no body or no experience at all. In short, impure subjectivity is a form of subjectivity that is not based on identity, rather on difference. As I will explain in the last part of this essay, it is because of this persistent difference that imaging technology of the inner body will never succeed in representing the *living* body. Secondly, bodily subjectivity means that consciousness cannot be something enclosed upon itself; it is embodied and embedded. Hence, it is not such a secret and intimate place as we thought at first. It is true that someone else might not feel the same joy as I do when I drink my coffee, but it does not result from this that these so-called subjective and inner feelings originate from a private and enclosed inner self. The very possibility of having these kind of feelings presupposes that one must have been touched, in one way or another, from the outside.[10] In what follows, I will explain these two lines of reasoning—the subject as difference and the subject without privacy—on the basis of Jean-Luc Nancy's reflections on the body. One of the central themes in his work is the debunking of the idea that something like subjectivity or interiority, i.e. something like the 'inner' feeling or experience of oneself, is separated from anything outside. According to him, interiority or subjectivity is always related to the tangibility of the 'outsideness', hence his idea of *ex*teriority.

Exteriority: Sensing Oneself from Outside

Subjectivity or interiority, in the psychological sense of the word, implies identity. Indeed, as I explained earlier, the subject is the being that remains the same, identical; it is the underlying substrate that remains the same whereas other features may change. It is a prevailing conviction that one's inner self forms one's identity. Identity in its strict sense—numerical identity—implies also unicity: I am I and not another. Identity is thus based on the idea of an internal *adequatio* between I and I (I = I). Yet, if the experience of one self is based on the difference between *Leib* and *Körper*, the very idea of an *adequatio* is put into question. The difference between *Leib* and *Körper* is, in fact, the difference between sensing and being sensed, or touching and being touched. Although he never mentions the distinction between *Leib* and *Körper*, I would say that the difference between touching and being touched—*la touche*—forms the central theme in Nancy's idea of the body. To name the body's 'ambiguity', its inherent difference, he rather provocatively conjures up the notion of the soul. He claims that the soul is the body which is related to itself. It is the body that feels itself (*se sentir*), and as such it is the body that does not coincide with itself,

[10] The demystification of interiority I am aiming at here bears some resemblance with Wittgenstein's analysis of (inner) feelings. According to him, a so-called private feeling such as pain is not private at all; for, it already presupposes a language in which this feeling is expressed, and language is never private but rather a social practice. (Wittgenstein 1953).

the body that is *outside* itself.[11] The term soul is used to make clear that the body is not a thing, a something, but rather a *relation*. What is thus at stake in Nancy's work is the attempt to explain the relation, or difference that constitutes (bodily) experience. This relation is prior to the sides or the aspects which are related. It is for this reason that Nancy does not talk about the different 'aspects' of the body such as *Leib* and *Körper*. The body should not be considered from the angle of its double-sided character, but from within its difference, its relation. This paper is not the place to discuss the complexity of such a philosophical approach. For our purpose, it suffices to underline that according to this line of reasoning, the body is not from the order of the thing or the substance. It is from another order—an order that can do justice to the being of the relation—which can be named the order of the *subject* (Nancy 2006a, 109).

Substance and subject must not be confused.[12] Substance refers to everything—be it material or spiritual—that is enclosed in itself and that coincides with itself. Substance is mass. It is mass without extension, without exposition. If a body is considered to be a substance, instead of a subject, it is reduced to a mass, or as Nancy argues with a reference to the cruelties of the Bosnian war in 1994, it is reduced to the mere materiality of a mass grave. As such, it is no longer a body that can be genuinely touched. The body as mass or as part of a mass is in fact nothing else than the denial of its 'being body' (Nancy 2006a, 107). By contrast, the body as subject can be touched, by itself and by

[11] "The soul is the difference between the body and itself. It is the relation from the outside which constitutes the body's being for itself"—"L'âme est la différence du corps à lui-même, le rapport du dehors qu'un corps est pour lui-même" (Nancy 2006a, 119).

[12] The distinction between subject and substance is a recurrent theme in Nancy's work. Especially in his reading of Descartes, he proposes to redefine the *ego* in terms of subject (Nancy 1979, 2006b). Subjectivity stands for openness, comparable with the Heideggerian idea of *Dasein*. If subjectivity is not based on substance, we need to revise the common opinion of identity. Normally, identity refers to the sameness of an underlying substrate, be it spiritual, mental or physical. If such a substrate is not the basis of a subject, the identity that forms the subject cannot be thought of in terms of simple *idem*-identity. It is therefore that Paul Ricoeur, also inspired by Heidegger's description of the *Dasein*, claims that identity of human existence—its endurance in time—is not carried by a substance but rather by a subject and this can be called *ipse*-identity (Ricoeur 1991). This is a form of identity that allows change and difference. According to Nancy, subjectivity or identity can only be understood on the basis of difference, as he makes very clear in the analysis of his own heart transplantation (Nancy 2000).

others, and in that way it is the body that belongs to a crowd (*foule*).[13] It is something open, not closed. When Nancy chooses to use the term 'soul' to indicate this opening or openness, one might get the impression that experience of the body—the fact of being touched—is now defined in terms of (psychological) interiority. This, however, is not the right conclusion. Nancy makes it very explicit that the soul should not be understood in a Platonian or Christian way. It is not the 'ineffaceable interiority' (Nancy 2006a, 113); it is not some spiritual body, since any spiritual body is nothing but a substance and therefore not of the order of a body that can be touched. In fact, something like a spirit is from the same order as shit. "Shit and spirit are the excretions of the body, that what the body rejects" (Nancy 2006a, 114). Even if I touch myself, this touch does neither presuppose nor constitute a self-coinciding interiority. Touch always comes from the outside and the touched always remains outside. The body's openness, its non-coinciding relation with itself, must therefore be understood in terms of *exteriority*.

One might be inclined to see exteriority as the opposite of interiority. However, according to Nancy, interiority is nothing other than exteriority. As explained earlier, we normally describe our 'inner' feelings, the (bodily) experience of our selves as (psychological) interiority. By using the term exteriority, Nancy claims that this current description is wrong. Our most 'private' feelings are not somewhere inside; they are at the outside.[14] Pain and joy are normally seen as the most private and intimate feeling. And yet, if I say that 'I suffer' or 'I enjoy' (*je jouis*), this 'I' is not a self-coinciding intimacy. In fact, these experiences already imply two 'I's'; in the case of suffering, one I rejects the other, and in the case of joy, one I exceeds the other (Nancy 2000, 39). Nancy does not explain any further what he exactly means with these two 'I's' which do not coincide, although "they are as like as two peas in a pod" (*deux gouttes d'eau*) (Nancy 2000, 39). I would say that this

[13] Central in Nancy's work is the theme of community, i.e. what it means to life with others, being part of a crowd. According to him, community should be understood on the basis of an ontology of 'being with' (*être avec*), and such an ontology must be an ontology of bodies: 'L'ontologie de l'être-avec est une ontologie des corps, de tous les corps, inanimés, animés, sentants, parlants, pensants, pesants' (Nancy 1996, 107).

[14] One might want to object here that we can have an experience of a certain interiority in our body, for instance, in the case of complete health. As claims Bichat, the definition of health is the genuine intimacy of a silent body (Nancy 2006a, 118). Nancy does not disagree with this idea of health, but for him this intimacy is from the order of mass or substance. It is not felt and as such it is not an experience of the body. In the case of complete health, there is no body.

non-coinciding relation can be understood as the relation between *Leib* and *Körper*. To be able to enjoy or to suffer one needs to have both a *Körper* that can be touched and a *Leib* that can sense its being touched. As I see it, it is precisely the ineradicable persistence of the *Körper* that disrupts the possibility of an interiority. Indeed, it is the *Körper* that can never be internalized, that always remains outside. Exteriority then stands for the seemingly paradoxical fact that I can only experience my own body on the basis of something that is and remains strange; the strangeness of the body's thinghood, its *körperlichkeit*. If we summarize this philosophical account on interiority we can reach the following argumentation: (a) interiority has not only a spatial but also a psychological meaning; (b) psychological interiority stands primarily for the inner, private feeling or experience of one self; (c) this so-called interiority is, in reality, always based on something *exterior*, something with *extension*, something that cannot be internalized or appropriated; (d) things with extension (*Körper*) are spatial. So-called psychological interiority, that we now have reformulated as exteriority, is thus related to spatiality. And in principle, everything spatial can be represented by any kind of imaging.

The inner space of the living body is unveiled by contemporary imaging technologies. In the introduction, I made already clear that these kind of images will never unveil our so-called 'inner feelings'. But since we have reconsidered this interiority in terms of exteriority—which always remains related to some kind of extension and thus spatiality—we need to explore in which way these images touch upon the body's exteriority.

What is on the Screen?

Various imaging machines can make images of my interior body, which I can look at myself—and sometimes, even in real time (ultrasound, endoscopy). This means, in fact, that my body is looking at my own body: the body is simultaneously seeing and been seen. While describing the phenomenon of bodily subjectivity, based on the difference between *Leib* and *Körper*, I used Husserl's analysis of touch. The touched hand is not merely touched or sensed, rather it senses its very own touchability. There is a reversibility between sensing and being sensed, which at first sight might look typical for the sense of touch. Merleau-Ponty, however, claims that this model of touch equally counts for vision. As he writes, vision is a 'remarkable variant' of tactile palpation (Merleau-Ponty 1964, 133). Consequently, seeing implies always being seen. Here, I will not elaborate on this explanation of vision that is modelled after the reversibility and proximity between touching and being touched.[15] If we pursue Merleau-Ponty's account, we may say that it is also by means of vision that we can experience our body both as *Leib* and *Körper*. The *Körper* is the body that is seen, and the *Leib* is the body that is seeing. My body is not only a thing that can be seen but it is also seeing. It is a *Leib* because this 'seeing', while entangled with movement and space, is not a (Cartesian) mental way of seeing, but rather an embodied seeing.

The experience provided by medical images of our own body seems to affirm this double-layered experience of our own body; seeing our own body with our own body. However, there is also an important difference to mention. It is the machine that constitutes the difference between the two layers of my bodily existence. There is no intrinsic and immediate relation between the body that sees and the body that is seen. Moreover, the relation between seeing and being seen is complicated by the fact that we as laypeople in general do not immediately recognize what we see.[16] The machine thus creates the difference between *Leib* and *Körper*. Or to phrase it more precisely, it creates a distinction between, on the one hand, a pure *Körper* and, on the other, a *Leib* that is also *Körper*. The image of the body always represents the body as *Körper*, as a thing, an object. Indeed we cannot make a physical image of something that is not extended. How does this image of the body as *Körper* relate to the notions of interiority and exteriority?

As I already suggested, the body as *Körper* is that dimension or aspect of the body that remains strange and exterior to the body as subject. The bodily subject cannot appropriate or assimilate its own *Körperlichkeit*. According to phenomenological vocabulary, the *Körper* is an intentional object, it is an object that can be perceived, an object with physical qualities. A *Körper* can be perceived as warm, smooth, hairy, slimy, smelly etc. In that sense, also qualities that are normally

[15] In another paper, I have provided a detailed analysis of touch and vision in Husserl and Merleau-Ponty (Slatman 2005b).

[16] If patients are frequently submitted to imaginary examinations they do learn how to read the pictures after a while. See for this Radstake (2007). Elsewhere, I have explained that although these imaging technologies do not provide pictures that resemble, it is still possible to recognize something. (Slatman 2007).

attributed to a living body are in fact nothing other than qualities of a thing. The *Leib*—which is the living body from a phenomenological perspective—, by contrast, has no physical qualities. Its only quality is its 'capacity to be stimulated' *Reizbarkeit* (Husserl 1989, 164), which is nothing else than the 'sensing' (*Empfindnis*) of its being sensed. Because of this radical difference, *Leib* and *Körper* cannot be reduced to each other. And this also means that the *Körper* always remains the *Leib*'s stranger. However, it is important to emphasize here that *Körper* and *Leib* do not refer to certain parts of a body. It is more adequate to say that they refer to the double-sided experience of the body.[17] In the example of the two touching hands, the same hand can be experienced as *Körper* as well as *Leib*. Moreover, in touch, and according to Merleau-Ponty, also in vision, the relation itself is reversible: the touching can be touched, the touched can be touching, and the seeing can be seen, the seen can be seeing. Obviously, this reversibility also reverses the relation between *Leib* and *Körper*. But again, this does not mean that the *Körper* can be reduced to the *Leib* or *vice versa*. As Merleau-Ponty made it very clear, there always remains a difference, an *écart* between the two. In the example of the two touching hands this *écart* consists in the fact that one and the same hand can never be touched and touching at the same time.[18] Although *Leib* and *Körper* are exchangeable, the difference, the *écart*, between them is persistent. And it is true, we could not even think of a bodily subject without this difference: if it were a pure *Körper* it would be just an object without an experience of itself, and if it were a pure *Leib* it would be something disembodied that could not be touched or seen. If our living body is the *Leib* that is always related to the *Körper*, and if images of the body can only exhibit qualities that belong to the body in the sense of being *Körper* what do these pictures tell us about our own body?

If the invasive camera or the scan enters the body, travels beneath the skin, it will exposure something that was first hidden and inaccessible.

---

[17] If *Körper* and *Leib* would denote some specific body parts, it would be incomprehensible that strange elements, for instance, prostheses or grafts could be appropriated. A *Körper*, or even better, a *Fremdkörper* can become part of my own living body and thus become *Leib*, but only to the extend that it ceases being *Körper*.

[18] "My left hand is always on the verge of touching my right hand touching the things, but I never reach coincidence; the coincidence eclipses at the moment of realization, and one of two things always occur: either my right hand really passes over to the rank of the touched, but then its hold on the world is interrupted, or it retains its hold on the world, but then I do not really touch it" (Merleau-Ponty 1964, 147–8).

Certainly, imaging technologies are capable of laying open a certain strangeness of the body. Or even stronger, they can even neutralize (or colonize) this strangeness. However, this is not the kind of strangeness that forms exteriority. As Nancy claims, exteriority is constituted by a stranger (*l'étranger*) who is not given a welcome, who in that sense always remains strange (Nancy 2000, 11). I would therefore say that the strangeness that is encountered by imaging technologies can have two faces. On the one hand, it refers to a strangeness that can be welcomed and thus neutralized. But on the other hand, they also face a strangeness that cannot be discounted. This strangeness is nothing other than the fact that the body in the picture is reduced to a *Körper*. A picture of the living body thus exhibits life's stranger. Instead of neutralizing strangeness, the invasive gaze only intensifies it. The machine increases the distance between *Leib* and *Körper* and externalizes their relation. The experience of one's own body as subject is always mediated by the facticity of the *Körper*, and the machine reinforces this mediation. In conclusion then, instead of unveiling the inner body's interiority, contemporary imaging technologies make visible the myth or illusion of interiority, by blowing up one's own living body's stranger—the *Körper*. It might be true that modern imaging technologies are so fascinating since they seem to make us transparent, since they appear so appropriate for understanding our genuine selves. I think however that the fascination for this kind of technology is first of all a fascination for what is not understandable at all. It is the fascination for our own death which apparently is so inextricably bound up with our life. Thomas Mann describes this fascination very aptly in his *Magic Mountain* (1924). Seeing his own chest X-ray, Hans Castrop has the impression of glimpsing in his own grave; a glimpse which is not at all appropriate for a human being.[19] A picture of the *Körper* which in fact belongs to a living body—one's own living body—separated from the *Leib*, is a picture of the living body's death.

[19] "Und Hans Castorp sah, was zu sehen er hatte erwarten müssen, was aber eigentlich dem Menschen zu sehen nicht bestimmt ist und wovan auch er niemals gedacht hatte, daβ ihm bestimmt sein könne, es zu sehen: er sah in sein eigenes Grab" (Mann 1924, 232).

CHAPTER SEVEN

LOOKING FOR A SPONGE: HOW A BODY LEARNS TO BE AFFECTED BY ULTRASOUND

Maud Radstake

> (W)e should be able to observe empirically how an anonymous and generic body is made to be a person: the more intense the shower of offers of subjectivities, the more interiority you get.
>
> (Latour 2005, 208)

It was at the desk of the Radiology Department that I first met Mr. Coenen.[1] He had an appointment for an ultrasound examination of his abdomen and kidneys. I told him that I was studying the role of medical images in people's experience of their body, their possible illness and its medical treatment and asked whether I could be present during the examination and interview him afterwards. Mr. Coenen agreed to that.

Fifteen minutes later, I was watching Mr. Coenen on an examination table in an ultrasound room. An assistant physician moves the ultrasound transducer over his lower abdomen. Turning his head, Mr. Coenen can see the monitor that the physician is looking at. As soon as the monitor shows the first grayscale images (fig. 7.1) Mr. Coenen asks: "*Are those the kidneys you see*?" "Yes", the radiologist confirms. "And all the black follicles are cysts." "*And the light pieces are what is still ok*?" "That is the kidney center." The screen shows moving pictures of Mr. Coenen's right kidney. The radiologist stores a still picture of a large cyst every now and then by pushing on a button on the ultrasound machine's control panel. "Let me look to the other side", she says. "*Should I turn?*" "No, just stay like this". The radiologist moves the transducer to the left side of the abdomen. She types in 'left' on the keyboard and the word appears next to the still image on the monitor.

[1] The story of Mr. Coenen is one of the cases documented in *Visions of illness. An endography of real-time medical imaging* (Radstake 2007), which is based on fieldwork in a radiology and in an endoscopy department of a Dutch hospital.

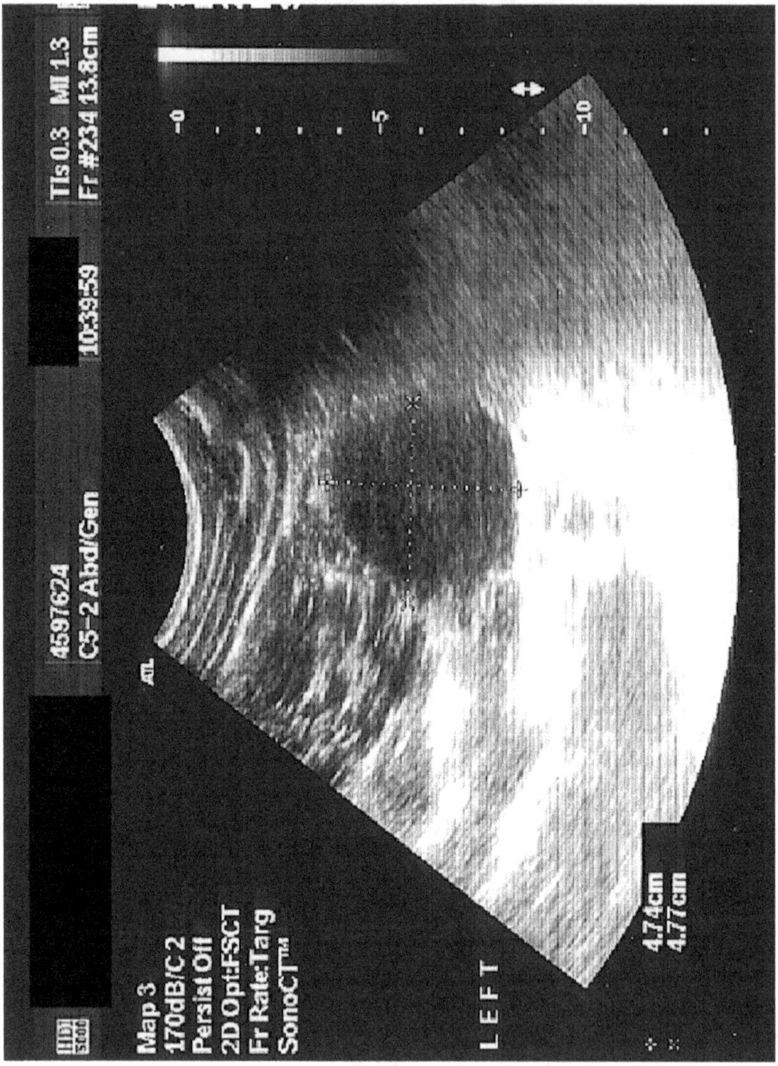

Fig. 7.1. Ultrasound picture of right kidney. Reproduced with permission of the patient and the radiologist.

"*Is it worse than the other side?*" Mr. Coenen asks. "Hard to say, it is difficult to quantify. Can you hold your breath for a moment?"

The radiologist uses the mouse to move a pointer across the screen. She clicks it to freeze an image and marks the outline of a black circle (fig. 7.2). "*What is this?*" "Another cyst". "*A big one, isn't it?*" The radiologist measures the cyst with another mouse-click, its size appears on the monitor. It is indeed the largest cyst measured during this examination. The radiologist moves the transducer to show the liver. "I also see some cysts here", she says. "Don't worry, that's quite normal. When you have a disposition for cysts, you can have them anywhere. There are not many and you don't have any complaints, do you?" "*No, I haven't*". "Alright then, that was it. Goodbye."

Translating ultrasound images

Ultrasound is a medical imaging technology that enables not only medical professionals to watch the interior of a living human body in real-time, but patients as well. By training and experience, the radiologist had learned the specific grammar of moving ultrasound images: how to produce a meaningful image of a kidney, how to identify a black circle as a large cyst, how to measure it and compare its appearance and size to normal images and values, and which images to store for future reference or just for the patient's record. The patient in the above case was neither trained nor experienced in watching those images. Mr. Coenen asked questions to understand what he saw. Thus he participated in the radiologist's reading of the moving pictures. He participated in the making of the images as well: by lying on the table with his upper body undressed and by answering to the radiologist's calls to hold his breath or breathe normally again. While engaged in all those activities Mr. Coenen felt the transducer moving and watched the shifting shapes, words and number on the monitor.

After the examination the radiologist went straight to a nearby dictation machine and started dictating the report for the examination: "Examination in preparation of kidney transplant, to check size of polycystic kidneys and presence of stones in gallbladder and urological tract. Both kidneys contain a pronounced quantity of cysts of various sizes…". She did not retrieve any of the pictures from the digital system containing the stored still images, but reported from memory and from the few notes she had taken during the examination.

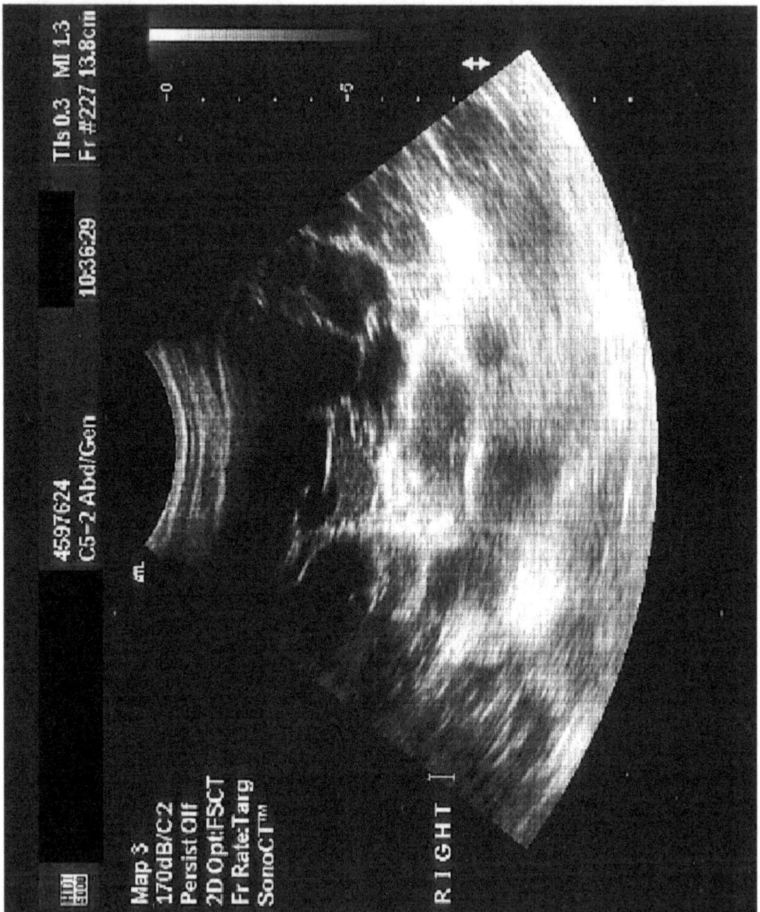

Fig. 7.2. Ultrasound picture of large cyst in left kidney. Reproduced with permission of the patient and the radiologist.

Meanwhile Mr. Coenen got dressed. I met him a few instants later in a nearby empty examination room, where we talked about the examination and about his illness. Mr. Coenen recounted that he had known for twenty years that he had polycystic kidney disease. Several relatives suffer from the same disease. Since the diagnosis he has visited the hospital regularly for check-ups. Recently he has suffered from increasing hypertension and fatigue. Those are signs of diminished kidney functioning. The specialist attending Mr. Coenen requested today's ultrasound examination to diagnose the condition of the kidneys. The examination has been one in a series of examinations with the objective of assessing whether Mr. Coenen should start with kidney dialysis soon, awaiting transplantation.

When I asked Mr. Coenen after the ultrasound examination what exactly was the condition he suffered from, he told me how his attending physician had explained it to him: "*A kind of sponge he said, with small holes.*" When I asked him whether his kidneys on the ultrasound indeed looked like what he expected, Mr. Coenen replied: "*It looked much worse than I imagined. The holes were much bigger than I expected. Actually I am a bit shocked. And then in the liver, she found some cysts there as well, I did not know about it.*"

Watching pictures, seeing illness

Having an ultrasound of his kidneys made Mr. Coenen more ill, in two related senses. First, the images showed him that his condition was worse than he had thought, since his kidneys did not look like a sponge. That is, the images affected his experience of being ill—what sociologists and anthropologists of health and medicines have called *illness*. Second, the images showed *disease*: the radiologist's reading of Mr. Coenen's physical condition.[2] A week after the ultrasound examination Mr. Coenen went to see his attending specialist, who told him that the ultrasound combined with the various other examinations indicated that his kidneys were in such bad shape that time had come to start dialysis, probably within weeks.

[2] The illness/disease distinction is central to medical anthropology, a sub-discipline of medical anthropology (cf. Eisenberg 1977; Kleinman 1980 & 1988; Helman 2001).

Mr. Coenen and the radiologist read the ultrasound images differently. The radiologist was not so much concerned with the number and size of the cysts but rather with the size of the kidneys, which were actually too large to fit the ultrasound screen. Mr. Coenen was not really impressed by that finding. His attending physician had already told him after a physical examination that his kidneys were very large. What alarmed Mr. Coenen was that his kidneys did not look like a sponge. However, Mr. Coenen's reading was informed by the radiologist's real-time interpretations as well. Despite initial worries, he was easily reassured about the cysts found in his liver.

Mr. Coenen watched the images of his kidneys as well as those of his liver. The impact of the pictures, however, was very different. Mr. Coenen explained: "*You see a liver, but actually I cannot really imagine it. She just tells me: this is the liver. It is easy to recognize the kidneys. Well, not really recognize, but you know that it is the thing with the holes in it.*" For the most part, visceral organs or phenomena[3] are unavailable to people's conscious awareness and control. One can directly perceive one's exterior, yet visceral perception is largely indirect and requires mediation of some sort, often via the surface of the body. American physician and philosopher Drew Leder has written a phenomenology of 'the absent body', distinguishing different ways in which the body can escape experience (Leder 1990). He describes the body as a field "in which certain organs and abilities come to prominence while others recede" (Leder 1990, 24). Liver, kidneys and other organs of the interior body are hardly present as objects of conscious experience. Yet for Mr. Coenen, his kidneys did feature prominently in his experience of his body. For watching his liver, however, Mr. Coenen lacked the framework that guided his view of the images of his kidneys: the image of the sponge.

During the examination, the body of Mr. Coenen was present as an object of sensory perception, as an image and as an actor. In a real-time imaging event, all those modalities of the body coincide. The images are momentary compositions: they only exist while they are made, as composites of heterogeneous elements, including the hardware and software of a technological imaging device; a doctor and a patient, each

[3] "The category of the "visceral" understood broadly includes not only the organs of the digestive system but of the respiratory, cardiovascular, urogenital, and endocrine systems, along with the spleen" (Leder 1990, 37).

with her or his own particular behavior; sensory perceptions; request forms and protocols; physical body parts; and a particular time-frame. The radiologist digitally stored some stills of Mr. Coenen's examination, but those were disconnected from the actual time and place of production. Characteristic for ultrasound and other real-time medical imaging technologies is that the images are largely read on the spot, in the presence of the patient. What real-time images show, results from a specific configuration of elements at a particular point in space and time.

A lot of work by physicians as well as by patients and machines is required in order to configure a body that can be imaged. In real-time, there is no doubt about the referent of the images: during an examination, neither the physician nor the patient questions whether the pictures on the monitor represent the body on the examination table. Both to the radiologist and to Mr. Coenen the ultrasound images showed that something was the matter. Yet the nature of that matter was different. The radiologist looked for the size of the kidneys and combined the visual data with the results of other examinations into information that was relevant in view of anticipated dialysis and transplantation. Mr. Coenen understood how ill he actually was by comparing the size and amount of the kidneys to the image of a sponge.

In the preceding phrases, visual perception and interpretation are attributed to a singular actor—either physician or patient—whose perception is mediated by a machine. Yet in actual practice, the ability to watch images is distributed among a network connecting body parts, an ultrasound machine, a trained physician and a patient on an examination table. Before the examination, both the object—the body watched—and the subject—the body watching—are anticipated in temporary constellations of people and things. During the imaging event, various modes of embodiment are enacted. There was the body that Mr. Coenen sees, feels and acts with; the body on the examination table that was the ultimate object of medical examination; and the body on the monitor as black and white shapes or colorful structures. A real-time imaging event presents different modes of embodiment at the same time.

Different modes of being and having a body, resulting from particular practices of embodiment, give rise to particular experiences of one's own body. Those experiences do not merely consist of interpretations, but they also of the material, perceptual impact of those phenomena on one's body and senses. That impact has been labeled as 'presence'

(Gumbrecht 2004). Besides an interpretation of meanings, any account of experience should address how objects are presented in such a way that they initiate or intensify a tangible, immediate impact on human bodies and their senses (Gumbrecht 2004, xiii).

So far I have presented Mr. Coenen's experience of the examination and of his illness largely through his own narrative. A narrative is a condition for experience rather than a representation of it (Desjarlais 1996). And although experiences are obviously shaped by the ways in which they are expressed, they cannot be reduced to their expressions. Inspired by phenomenologists, social scientists have stated that illness narratives constitute experience (Hydén 1997). Focusing on narrative representations, however, leaves the material and perceptual presence of bodies unattended. The work of anthropologist and philosopher Bruno Latour offers some conceptual tools for addressing such issues.

Illness as a circulating referent

Latour has introduced the concept of 'circulating reference' as alternative for a representationalist epistemology (Latour 1999, 24–79). The idea is that reference is not a relation between an object and representation, but comes into being in a series of material and semiotic transformations of both. An object of knowledge, for instance a patient's body in a diagnostic practice does not precede the real-time images that supposedly represent it. Referent and representation are linked in a series of translations. The ultrasound pictures appearing on the monitor before Mr. Coenen do not resemble his kidneys nor do they depict his medical condition. The phenomenon that "*circulates* all along the reversible chain of transformations, at each step losing some properties to gain others" (Latour 1999, 71, emphasis in original) in the examination I have described was polycystic kidney disease, the illness of Mr. Coenen.

When looking at the practice of real-time medical imaging, the distinction between *illness* as the object of patients' experience and *disease* as the object of professional medical knowledge production is not self-evident. The radiologist's readings of the ultrasound pictures, while partly evoked by Mr. Coenen's questions, also informed the patient's spectatorship. Mr. Coenen's cooperation was a condition for the production of readable images that in their turn opened up his body for medical scrutiny and visual experience. Real-time imaging entails three connected chains of translation. One is the transformation of a body into an object for medical visualization and clinical knowledge production. Then there is the translation from body to images. Simultaneously, an embodied patient becomes co-producer and spectator of those images. Those processes of objectivation and subjectivation are connected in one body and in one particular moment in time. The referent of real-time imaging is both illness and disease. For my purpose of understanding the experience of Mr. Coenen, I prefer to use the word illness. However, this notion of illness includes the radiologist's readings and the ultrasound images.

The scheme in figure 7.3 presents real-time imaging as a chain of translations (cf. Latour 1999, 70). The different stages of translation are not directly linked, avoiding the suggestion of univocal relations. At each step in the chain, the referent of medical imaging is materially and discursively adjusted to the requirements of a particular configuration of actors. The dotted lines represent co-enactments of embodied objectivity and embodied subjectivity. The referent circulating along this chain is *illness*: the potential object of medical examination and of embodied experience. The illness of Mr. Coenen resulted from the real-time co-presence of his body and the images on the monitor. His body got connected to the images before, during and after the imaging event.

The trajectory of Mr. Coenen's ultrasound started with the preparations for the examination. The physician attending Mr. Coenen based his decision to refer him for an ultrasound on his knowledge about the patient's condition and about the ability of ultrasound to visualize the suspected condition. The ultrasound machine was configured for an examination of the kidneys and abdomen by a radiology assistant right before the examination. The radiologist performing the examination had been trained to produce and read real-time images. Mr. Coenen

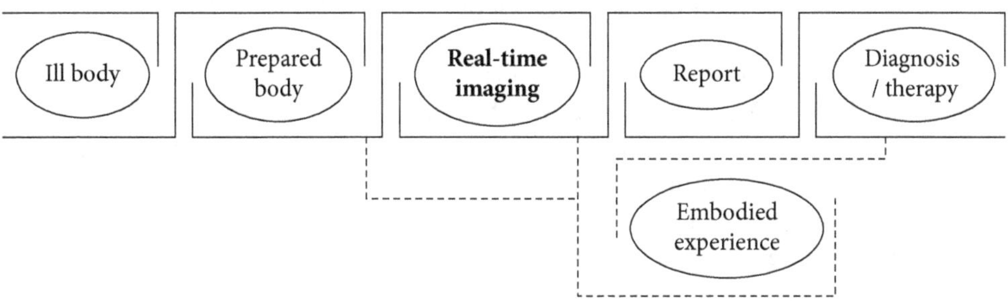

Fig. 7.3. Schematic representation of the stages of bodily translations in the chain of real-time medical imaging.

himself read the information provided by the radiology department and adhered to the instruction not to eat or drink any gas-inducing products two days prior to the examination.

In Latourian terms, Mr. Coenen is enrolled into a network of real-time imaging, because his physician requires images in order to decide upon future medical action. In that process Mr. Coenen is transformed into an object and into an instrument of medical imaging that needs to be prepared in order to play those roles. The agency to show and watch real-time images is a composed property of the network (Latour 1999, 174–215). All preparations focus on the patient's body as an object for visualization. Yet the body on the examination table is more than an object to be visualized. It was also Mr. Coenen, very much aware of his malfunctioning kidneys, who had actively prepared himself and who was able to watch the live moving pictures on a monitor. The information brochure did not mention that possibility.

During a real-time imaging event, a patient's body and the moving pictures on the monitor get connected in various ways. A first attachment is established by the radiologist sharing her real-time readings with Mr. Coenen. The simultaneity of Mr. Coenen watching the images on the monitor and feeling the machine that produces the images moving over his body constituted a second mode of attachment. Thirdly, Mr. Coenen was linked to the images by actively participating in their production, for example by holding his breath when the radiologist asked him to. The simultaneous real-time presence of his body as an image and as his own lived body constitutes an undeniable agential connection between Mr. Coenen and the ultrasound pictures. Michel Callon and Vololona Rabeharisoa have described something similar for people with muscular dystrophy who watch their own bodies in a mirror. The visual objectivations transform the ways in which their body is able to perceive and act as a result of the constant going back and forth between being a body and having a body: "(i)t is the miracle of the mirror which makes these two experiences possible at the same time and makes this essential tension visible" (Callon and Rabeharisoa n.d.). Finally, Mr. Coenen was also emotionally affected by the images, because they showed him his body and his illness, which worried him. Those four modes of attachment—hermeneutic, perceptive, agential and emotional—made Mr. Coenen watch *himself* on the monitor. His body became transparent, his illness visible.

Real-time imaging presents the interior body as an image, as something that is the object of sensory perception informed by cultural notions. This is what philosopher of technology Don Ihde has called an image-body. The image-body differs from the here-body, which is the body that perceives it (Ihde 2002, 5). During real-time imaging, here-body and image-body are connected. After an examination, however, the real-time connection of referent (body) and representation (real-time images) is broken. They get reconnected in various expressions of experience. While physicians produce a referent by making a report of the imaging event, patients take their memories of the examination home with them and translate them into stories told to relatives, friends or an inquisitive anthropologist like myself. It was not self-evident that the ultrasound images came to represent the body and illness of Mr. Coenen. It required a lot of work by all actors involved: the radiologist, Mr. Coenen and the ultrasound machine.

Physicians' and patients' translations of real-time imaging events result in medical reports and illness narratives presenting objectified images and bodies. Four kinds of image-bodies are produced in the translation of real-time imaging events to embodied objects of patients' knowledge and experience. The four image-bodies constructed as referents of real-time imaging events are not simply the result of the four modes of attachments constituted during an examination. Image-bodies result from the translations of real-time images after they have ceased to exist.

The *medical body* is one of the image-bodies. Constructed in the professional readings and translations of body and images during and after the imaging event, it is the body that is presented in the imaging report. The medical body is the referent of an assembly of words and numbers that refer to images as visual data. The *transparent body* establishes a second type. It is the result of increasing entanglements of medical and media technologies (Van Dijck 2005). With their transfer to the public domain through newspapers, magazines and television shows, medical images have become powerful epistemological devices by which people understand the world (cf. Sontag 1991). The transparent body is a cultural construct rather than a medical ideal. The complexity of medical imaging production and interpretations tends to get lost in media translations, as was illustrated by some patients I spoke to, who compared real-time medical imaging to live television. A third type of referent is the spectacle that the patient actually watches on the monitor, the *spectacular body*. Finally, real-time images refer to one's *lived body*, actively engaged in the imaging event as embodied actor and spectator.

Learning to be affected

The above analysis of real-time imaging producing medical reference as well as patient's agency could be summarized by the notion of mediations. Real-time imaging technologies mediate between patients as embodied subjects and their bodies as objects. They offer patients particular possibilities to appropriate, or incorporate, material and perceptual features of their bodies that have been produced in their connections with others. Bruno Latour has used the concept of mediation to describe the material-discursive enactment of agency. For others, however, the concept has a hermeneutic connotation. Therefore, I prefer another expression coined by Latour to describe the work of real-time imaging in the formation of embodied experience: that of *bodies learning to be affected.*

> When you enter into contact with hospitals, your 'rich subjective personality' is not reduced to a mere package of objective meat: on the contrary, you are now *learning to be affected* by masses of agencies hitherto unknown not only to you, but also to doctors, nurses, administration, biologists, researchers who add to your poor inarticulate body complete sets of new instruments (....) No subjectivity, no introspection, no native feeling can be any match for the fabulous proliferation of affects and effects that a body learns when being processed by a hospital (...).
>
> (Latour 2004, 227, italics MR).

Real-time imaging technologies trigger a series of translations of bodies and images into image-bodies, which function as objects of patients' embodied experience. Rather than being the material basis for the formation of an embodied identity, one's body is "an interface that becomes more and more describable as it learns to be affected by more and more elements" (Latour 2004, 206). The imaging devices, monitors, pictures, numbers, gestures, spoken words, reports, cultural images and sensory perceptions involved in a real-time imaging event are examples of such elements.

The possibility for patients to watch the monitor during an examination makes it possible that medical images affect the patient as an embodied being, because he or she can relate to the objectified body as an actor.[4] Watching the monitor during an examination does not provide patients with a meaningful ground for agency and identification

[4] This phenomenon has also been described as *synecdoche* (Cussins 1998).

just like that. When they lack a frame of reference, patients cannot link the represented body to their embodied self. Mr. Coenen attentively watched the monitor during the ultrasound examination. Many other people whom I observed did not, either because they lack proper guidance or because they are happy to leave the images with the doctors. Those patients did not constitute a perceptual link between their bodies and the images on the monitor. That link may not be established for other reasons as well, for instance when the doctor's findings contradict the patient's physical sensations. Philosophers and social scientists have described such experiences as alienation. According to such authors, technologies and practices of modern medicine constitute a move *away* from patients' own experiences and embodied selves (e.g. Duden 1993, Leder 1990, Toombs 1992).

In their analyses of medical practices, sociologists and anthropologists have frequently opposed objective medical conditions that "can be directly read from the body or discovered through laboratory tests" (Baszanger 1998, 145) to experiences of pain and illness that are private, personal and only indirectly accessible through patients' narratives. Diagnostic technologies are then opposed to subjective expressions of experience, either explicitly in terms of alienation, or implicitly in terms of knowledge versus experience. However, alienation is no more a self-evident feature of real-time imaging for patients than a transparent body or a visible illness is for doctors. Whether real-time imaging renders illness visible or alienates a patient from her or his body is not determined by imaging technology, but results from an entire network of visualization that is formed by physicians, patients and technological devices before, during and after an imaging event. In that network, a body learns to be affected.

Inside out

The notion of embodiment as 'learning to be affected' denies any distinction between an 'inside' self and an 'outside' body. Patients are capable of being a self and acting as oneself through the multiplication of their connections with 'outside' elements. In Latour's words: "you need to subscribe to a lot of subjectifiers to become a subject" (Latour 2005, 216). Real-time images are neither outside representations that suppress embodied experience nor are they immediate self-evident views inside oneself. The enactment of image-bodies in patients' and professionals' translations of real-time imaging events is a process in which patients actively learn to be bodies.

In the case of Mr. Coenen the translation of real-time images—into a diagnosis by the physician and into a consistent narrative by Mr. Coenen—seemed unproblematic. However, the ultrasound did not simply show the physician that Mr. Coenen was due for dialysis. That decision required other examination and a careful assessment of all results. Nor did the ultrasound just show Mr. Coenen how he was doing. Him feeling more ill after the examination had to do with what the radiologist said, but also with the difference between what he himself saw on the monitor and the image of a sponge in his head.

Before, during and after the examination the ultrasound pictures set up the body of Mr. Coenen and thus at the same time the illness that he had and the patient that he was. Rather than having or being a body, embodiment describes the process of becoming a body in a particular assembly of people, organs, machines and images. Real-time imaging technologies enable the production of a characteristic co-presence of image and body. Experiences of illness or health are enacted in translations of that co-presence into more or less stable referent-bodies. Real-time imaging technologies interfere in the space between a lived body and its representations. Whether real-time imaging contributes to an alienating experience of embodiment and illness or evokes patients' identification with particular image-bodies hinges on the (in)compatibility of physicians' and patients' translations in practice. The diffuse and contradictory experiences of patients taking part in real-time imaging events require a concept of illness that acknowledges and addresses the hermeneutic as well as the material and existential layers of embodiment. It is a concept of illness as a circulating referent and a mode of distributed embodiment.

The value of an ethnographic approach to the study of embodiment lies in tracing connections. Thus I found out that a different connection was established between Mr. Coenen and his kidneys and between him and his liver. The technology that visualized both organs did the same thing: sending high frequency sound waves which were reflected at different speeds depending on the density of the tissues and fluids in Mr. Coenen's body. It produced images of black circles in areas of light matter, which the radiologist translated as cysts. The images of the kidneys indicated an advanced stage of polycystic kidney disease for the physician and made Mr. Coenen realize how ill he actually was. They demarcated the onset of a new stage in Mr. Coenen's illness: that of dialysis and waiting for transplantation. The cysts in the liver were considered abnormal as well, yet did not evoke any further action by either a physician or Mr. Coenen himself. Besides being mentioned in the imaging report, they did not leave any traces.

The translation of part of one's interior body into real-time medical images could commonly be described as a process of objectivation: an interior body that normally hides from perception comes forward as an object of knowledge and experience. We have seen how this exteriorization entails interiorization as well, being the transformation of subjective agency and experience. The crux of the argument, however, is that the production and translation of the simultaneous presence of body and image during real-time medical imaging constitutes a practice of embodiment that is neither exterior nor interior. Bodies are neither objects nor subjects, but interfaces.

Do real-time imaging technologies, with their possibility to show patients the inside of their bodies, present yet another instance of disembodiment (cf. Leder 1990, 98; Young 1989, 46) or do they contribute to the empowerment of the patient? Looking at situated practices, like the story of Mr. Coenen, shows the vanity of such questions based on dualisms that do not exist in practice. Watching ultrasound images of his own kidneys, Mr. Coenen's body did not become more or less his own. The images neither simply convinced him of a medical truth nor straightforwardly enabled him to take control over his illness. The results of an analysis like the one that I have provided here cannot easily be translated into practical recommendations What it can do, however, is contribute to a reflexive attitude building on embodied experience in actual clinical practice.

CHAPTER EIGHT

IMAGIN(IN)G PREGNANCY IN NORTHWEST TANZANIA: NETWORKS, EXPERIENCES, AND TRANSLATIONS

Babette Müller-Rockstroh

> Laurentsia Shitindi, 23 years old and living in Sengerema District in Northwest Tanzania told me about her first experience with a new technology she had when visiting the small district hospital with her third pregnancy that she thought did not grow as the others had: "How can you be fooled? The doctor looked at the kioo [Kiswahili: mirror/screen, the word health care professionals and women used to refer to ultrasound; BMR], the mzungu [white person; BMR] looked at the kioo. I saw the water, much water. All around the child, I saw it swimming there. How can I know that it is water? [...] He told me: 'This is the child there, there is the body of the child.' You could see it, the child was lying there on the left, and the rest was water only. [...] This new machine it can truly see each and everything."

In this chapter I try to trace the intertwinement of technological images of the pregnant body, with the hopes, expectations, fears, and the sense of one's own insides, in short of pregnant imaginations. I claim that both—ultrasound images and societal imaginations—are context-specific. More, I claim that context, content and the depicted objects—the body and that 'within the body'—are co-produced and are mutually shaping each other. In order to show this, I have followed ultrasound to a place where it has not yet become a standard, inevitable part of every pregnancy: I have followed ultrasound on its travel into the life worlds of women in Northwest Tanzania. Here, a number of hospitals have been rather recently equipped with ultrasound machines through a Dutch development project aiming to reduce maternal and child mortality statistics in the wake of the Millennium Goals, geared towards improving the lot and lives of all people.

Laurentsia Shitindi's story displays the fascination, shared by women in Northwest Tanzania and health care professionals likewise, for a machine that can "see each and everything". Her story at first glance also bears resemblance to the experiences of women with ultrasound in 'the West', where ultrasound examinations have turned into 'show-business' enjoyed by both expecting parents and professionals: "there is your baby; it is a girl; you see how she is yawning; and, doesn't she look like grandmother?" Yet, as I will show in the following, ultrasound in Northwest Tanzania and the bodies depicted by it, are but deceptively familiar. Studying ultrasound in contexts beyond Europe and the United States, in which this imaging technology was originally designed and put to use, I aim to contribute to a growing body of work that studies how bodies are being *done* in practice (cf. Mol and Law 2004) rather than imagine them to be the same. Studying the pregnant body as mediated, that is, as an effect of certain politics, histories, institutions and culture, however, requires not only to open the womb but technology itself to empirical scrutiny. Unblackboxing both ultrasound technology and the body is crucial for understanding pregnancy experience, analyzing questions of (dis)embodiment, discussing women's agency, and reflecting about the politics embodied in a technology that increasingly travels the globe, and by reaching far away places find new users and uses. Ultrasound may thus become a different thing in Tanzania, a low-income country where daily life is possibly more precarious and contingent, and political intervention in reproductive health more accidental than thoroughly planned. In turn, 'the body within' emerges as a situated set of practices and multiple genealogies.

Shooting the fetus: depicting women between alienation and agency

Over the past decades, the body reproductive has re-emerged as a focus of anthropological interest. A number of innovative studies have explored how reproduction is embedded in culture, in politics and in economy, in science and technology, in art and in ads. In particular how reproductive technology has shaped experiences of conception, pregnancy, and childbirth, and how notions of kinship and gender have therefore been challenged are at the center of these writings (see e.g. Ginsburg and Rapp 1995; Saetnan et al. 2000). Research on ultrasound in pregnancy has thereby been the focus of feminist and anthropological work ever since the technology became common in health care in 'the West'. During the 1980s, a large number of studies emerged from women's and cultural studies that presented ultrasound and other reproductive technology as factors favoring and reproducing male and technocratic usurpation of realms that had been women's stuff 'by nature' (see e.g. Petchesky 1987; Rothman 1993; Davis-Floyd 1992;

Duden 1993). Written in the middle of debates over abortion in the United States, and the congealing of doctor—and technology-centered delivery practices in American and European hospitals, this literature described women as alienated from their embodied experiences through the visual separation of mother and unborn by ultrasound that depicted the fetus as an independent free-floating entity—a separation difficult to overcome by the mother object/subject (see e.g. Young 1984). These studies provided an important cultural critique with regard to the imagined 'goods' of technology on women's lives. However, they often paid less attention to the actual experiences of women than to the imagined social consequences that the use of these reproductive technologies had for women's agency over their own bodies. Caught within the paradigm of 'technology as a Trojan horse', these critics took for granted that the political issues emerging from 'ultrasound in the West' were the same for 'ultrasound anywhere'.

Departing from the imagined community of 'sisters in ultrasound', recent studies of obstetric ultrasound ground themselves on the empirical study of women's experiences with this imaging technology, among these are Lisa Mitchell's research on 'baby's first pictures' in Canada (Mitchell 2001), Eugenia George's study in Greece (George 1996), Janelle Taylor's observations of the public fetus in the United States (Taylor 1992), and Ann Saetnan's work on pregnant women in Norway (Saetnan 2002). Rather than stressing women's disembodiment, these studies highlight women's agency as the effect of particular socio-culturally specific socio-technical configurations. Despite the richness of this research on reproductive technologies, however, it is striking to notice how few studies exist that empirically study reproductive technologies in low-income countries, that is, in cultural contexts that differ in many respects from the contexts these reproductive technologies were originally designed for and put to use. This gap in current research is particularly disconcerting as ultrasound—small, portable, comparatively cheap and modest in its infrastructural requirements—increasingly enters the 'world of flows' (Appadurai 1986, 3). Thus, we still know little about the encounters of ultrasound and pregnant women's bodies at these far away places.

It is only recently that the focus on 'ultrasound in Europe and the United States' has been complemented by studies such as that of Sigrid Tautz and colleagues about women's experiences with ultrasound in Botswana (Tautz 2000), Lilian Chazan's work about the commodification of ultrasound in Brazil (Chazan 2007), and Tine Gammeltoft's study on ultrasound imaging in urban Vietnam (Gammeltoft 2007). This chapter draws on their insights into the local bodies that come to meet with ultrasound, and thereby both affect and are affected by the new technology. Taking seriously feminist claims to study 'women' not as generic category but as poor, young, colored individuals, anthropology's claims to start from 'the native's point of view', and the credo of Science and Technology Studies that 'things could be otherwise', however, I want to push the boundaries that the lasting geographic focus on 'the West' had on thinking about the insides of the pregnant body even further. By 'shooting the fetus', that is by questioning the fetus as the inevitable 'natural' entity inhabiting the womb, I aim to leave the body open for different kinds of pregnancies to bear. I thus set out to study ultrasound worlds as culturally and historically specific socio-technical configurations that enact particular pregnancies, while they may exclude others. Based on interviews and participant observation carried out within my PhD research on 'ultrasound in Tanzania' in 2004 (Müller-Rockstroh 2007), the questions underlying this chapter are: How do ultrasound and pregnant women meet? How do they feed into each other? What are the objects that come to be depicted by ultrasound, and what normative value do these imaged facts carry? In short, how are pregnant/ultrasound bodies done in Northwest Tanzania, and which ones have or are granted the right to exist?[1]

Scanning pregnancy in Northwest Tanzania

In the health care facilities under study, ultrasound is not yet standard of every pregnancy. Pregnancy scans rather are the effect of careful clinical selection and subsequent referral that are shaped by the resources that exist and can be mobilized, by different kinds of expertise, and temporalities, and spatial conditions. Many women may thus still go through their reproductive lives without ever meeting with ultrasound. This chapter, however, is based on the narratives of women who have made it to the hospitals equipped with ultrasound, and whose reproductive trajectories have come to touch and be touched by the imaging technology enacting new images and imaginations. The five women

[1] This research was financed by NWO/WOTRO, and supervised by the Faculty of Arts and Social Science of Maastricht University, the Netherlands.

that will speak about their experiences with body scans are individual cases, yet they stand here as representatives for many other similar narratives that I have listened to throughout my fieldwork and that bear witness of the intricacies of technology transfer and the intention to introduce new bodies into existing practices to deal with pregnancy and pregnancy 'at risk'.

Contested unborns

When I met with Salima Kuku, 26 old and pregnant for the 13th time, she expressed her dissatisfactory experience with the new technology and the way it had been used by the nurses, and yet had not been able to see the agent responsible for the intrauterine death of all previous children—*nzoka*:

> I told them [the nurses] before. When I was young I used to eat earth. Maybe that is what is causing it. [...] My father said, it will give you worms but I was still young living with my stepmother. I was eating earth then. [...] She did not tell me not eat it, she was not my mother. She did not give me any medicine for diarrhea, for cleaning the stomach. This is how the nzoka came to sit in my stomach. Maybe it is worms, maybe it is nzoka. My first [and] my second child, it drank their blood, the children got lost in the stomach. My in-laws brought me to the mganga [Kiswahili: traditional healer; BMR], he gave me medicine for the nzoka to please it. He washed my stomach with the medicine, I heard it then. I heard its voice 'rrrrrrsh'. It came back in the next pregnancy. Then it moved here to the right, I heard it walking 'rrrrrrsh'. Now it hardens the stomach, so when the child grows and grows, it reaches here, it reaches the nzoka, it is bitten. The child dies."

As Salima Kuku's story illustrates, women's bodies in Northwest Tanzania are a shared place of humans and non-human entities. Quite explicitly Salima Kuku thereby refers to how people may acquire these entities—by eating earth, an abundant practice of pregnant women in the region around Lake Victoria, in Northwest Tanzania and Southwest Kenya (Geissler 2000). Yet, her story also mentions the more implicit causes for this shared body: disrespect for the orders of a parent, as well as carelessness of a stepmother to keeping a girl in order. Rather than separate, one being the biological 'natural' and the other being the 'cultural' cause, however, these practices are highly interrelated as they display the importance that earth has in agricultural societies. Earth is not only the ground for production and thus the basis for physical survival. It is also the ground in which to bury the ancestors and thus the basis for kinship relations. By eating earth for "refilling the blood" as I was told, women literally incorporate the lineage. Yet while living with nzoka thus is part of being a Msukuma,[2] somebody with a distinct and particular local identity, nzoka may also cause certain afflictions: it may provoke swelling of different body parts, it may cause people to fit, it may cover a woman's eggs with pus so that she cannot conceive anymore, they may cause the birth of abnormal babies, and it is known to drink the blood of mothers and children, thereby making mothers to feel dizzy and killing the child. The move from peaceful symbiosis to disease thereby hinges on gender and generation. Thus it is the unfinished state of the unborn that makes women-with-child more vulnerable to these kinds of afflictions than men, and it is a disrespectful child that may later suffer from nzoka, or a pregnant woman who disobeys her parents-in-law. Yet also the woman whose husband is "eating at other places" and thus may come home with "blood that is different" may be affected. What condenses in the term nzoka, hence, is a range of afflicting agents. These may originate both from biomedicine's hygiene practices and from local practices to providing the unborn and the mother with as much strength as possible, e.g. with necessary minerals inherent in soil (cf. Geissler et al. 1998), as well as they may originate from the transformations Northwest Tanzania has undergone in recent years. Thus, I was told, spirits had formerly lived in the big trees, yet had come to inhabit the bodies of people when these trees had been cut down for firewood for a growing population. All of these agents have one thing in common, though: they exert their power vertically, testing and contesting the unborn within the social fabric of life.

Moreover, Salima Kuku's narrative tells of the difficulty of diagnosing nzoka, and keeping it apart from worms. Making a correct diagnosis, however, is important given the fact that worms are entities to be treated by 'white' medicine, that is, by pharmaceutics. Nzoka, in contrast, is a matter for the traditional healer who can evoke the existence of this non-human entity by means of washing, thereby giving 'more body' to the relatively undefined entity 'woman with child'. Brought to the hospital by her husband to diagnose the cause for her habitual miscarriages, Salima Kuku's experience with ultrasound was one of resemblance: similar to the healer, also the ultrasound operator 'washed' her

[2] Msukuma is the singular form for a Sukuma person in Kiswahili, the plural is Wasukuma. The Sukuma tribe is the largest ethnic group in Tanzania and inhabits the southern shore of Lake Victoria in the Northwest of Tanzania.

belly with a liquid (the ultrasound gel) in order to see. Yet then, the operator failed to move "the eye of the machine" [the probe; BMR] to the site where Salima Kuku felt nzoka to be seated waiting craving for more blood and children, a hard texture shortly beneath her right rib that she made me touch with my hand.

Salima Kuku's account of nzoka displays common and shared knowledge of women in Northwest Tanzania. Whereas the diagnosis of nzoka in her story remained a matter of touch rather than of technology, other women told me with awe that indeed the new machine could see "each and everything", even nzoka. They thereby referred to what the nurses in antenatal care had told them after reading the ultrasound report: that the child was "lost in the stomach", and that surely the woman had seen the little white stripes on the screen? These were the "wadudu, the wadudu that had killed the child". For Waskuma women in rural Tanzania nzoka were wadudu in the sense of spirits, evoking a network consisting of ancestors, trees, societal misconduct, ethnic identity, etc. Nurses, however, used the term 'wadudu' in the second sense the word has in Kiswahili—small animals. Nzoka as wadudu here referred to a network consisting of syphilis, the main cause of the habitual miscarriages in the area, and thus a geographic identity, promiscuity, and individual misconduct. Building on (or perhaps also believing in) the omni-visibility of the technology/ body, ultrasound helped immensely to convince women such as Salima Kuku of the importance to follow syphilis treatment, said the nurses. In the translation work between languages and cultures, thus, a biomedical entity—ultrasound—here reinforced the existence of a traditional entity—nzoka—as a reality. Yet it simultaneously produced an entity that would (or potentially could) survive hospital practices and had therefore to be treated by traditional practices in order for a Msukuma woman to be able to carry children in the future. Hence, Salima Kuku told me, her next step after being discharged from the hospital was to go to the traditional healer and have him "scare nzoka again".

Concealed pregnancies

Mariamu Mpishi, 30 years old and pregnant for the fourth time, met with ultrasound because she had come "too late" to antenatal care, the nurse-midwife at the urban antenatal clinic scolded her, as to correctly assess gestation age. Pressed for an explanation for her delay after having carried three children already and thus being considered 'knowing',

Mariamu Mpishi revealed that she had been uncertain whether this was a real or a fake pregnancy:

> Our grandmothers used to tell us that people [...] that people in the past were being bewitched by given medicine through uchawi [Kiswahili: witchcraft/sorcery; BMR]. You would see someone with a big stomach but nothing would be born. Now I started thinking about my stomach which had become too big: "If it is not water, then what is it?"[...] Others say that there are people who give birth to smoke, others are bewitched with traditional medicine even now, in fact, my mind went very far.

Mariamu Mpishi's account was one of many stories that depicted another agent peopling and affecting pregnant bodies: other people's jealousy that would cause these people to use witchcraft in order to harm the obvious prosperity of their neighbors. In our interview Mariamu thereby pointed to her particular living conditions as having caused her imagining witches to be responsible for a belly "too big" for the number of months she had not seen menstrual bleeding. Having moved to the city where she worked as a cook selling food on the street, she and her husband, a fisherman, lived up the hills of town, a crowded area with one small hut next to each other. Urban living conditions where strangers live next door thereby differ immensely from the traditional way Wasukuma live, namely in several homestead inhabited by kin and spread out over a larger area. It was this living condition, she said, that may have made "someone want something", and carrying out this jealousy by bewitching her. Witches, as I learned, could thereby mimic a real pregnancy while in fact a woman would be only carrying stones or ashes imbued with movement 'as if' a woman was carrying a child.

Yet, not only could witches confuse a woman about whether being pregnant with child or with a fake child, they could also conceal a pregnancy by turning it to a woman's back. Unlike pregnancies turned to the back by women themselves, for example, in order to make a child born in the presence rather than in the absence of a husband "away for a very long period" that could easily be turned to the front again, pregnancies turned to the back by witchcraft resulted in malformed children with "twisted legs", "large heads" and bodies "covered with body hair all over".

The possibility to dislocate a pregnancy for an unlimited amount of time to a place where it can remain but does not grow, that is allotted to both women and witches, speak at once of women's possibilities to declare the social legitimacy of a child (that may be the product of an adulterous relationship), as it provides an explanation for obstetric pathology despite all preventive measures women take against witches. Practices to prevent affliction by witchcraft, for example, consist of drying underwear inside the house only, hiding and secretly washing menstrual cloths, by wearing necklaces with traditional medicine, or by referring to pregnancy in public only indirectly, by using the expression 'I have gained weight' [Kiswahili: nimepata uzito] instead of 'I am pregnant'.

Affliction by witchcraft requires the help of a traditional healer or diviner. Hence, all women I spoke to agreed that such a pregnancy would not be taken to the hospital, unless women were (made) uncertain about the agent of affliction. In Mariamu's case, for example, she was finally asked by a colleague what she believed in, and upon her answer that she was a Christian, advised her to go to the hospital.

Pregnancies that have 'turned to the back', could not be seen by the new machine, women said. Yet it was less due to the fact that these pregnancies were not taken to the hospital anyway, that accounted for this technological inability, but the fact—as women explained to me—that the machine was used only on the belly, and not on the back. My careful question whether this kind of pregnancy could be seen if the transducer was moved to the back of a woman was met with laughter: it could of course not, or had I ever seen ultrasound being used at the back? Yet, what if a pregnancy that had turned to the back by a woman herself would have turned to the front again? In this case, women told me, the machine would again work and see what the case was. I met Frida Mgalla whose family claimed had successfully turned the pregnancy to the front again and was now attending antenatal care as any other pregnant woman. Her mother then told me the following:

> You see this pregnancy was turned to the back. Why? It started the wrong time, so by the time the child would be born, we [would have] no money, no transport. The man was away. He came back only much later, so that child was to wait for some time. He is back now. He works in the mines. He is gone a long time. We went to the mganga, he gave us medicine. We turned it around again.

Admitted to the gynecological ward with the ultrasound diagnosis of a huge intrauterine mass, Frida Mgalla was discharged a week later to die at home with an inoperable tumor. Other than in the case of affliction by nzoka, the experiences of Mariamu Mpishi or Frida Mgalla with the new technology show, that in the case of witchcraft affliction ultrasound does not touch the network of traditional orders. Pregnancies that have turned to the back remain invisible to the pervasive technological gaze that people attribute to it—this invisibility being the result of its usage on a particular body part and not on another. Women's carry-ability, to be 'with child', may thereby remain untouched by a technology that potentially could prove women to be infertile or be barren, and carrying a tumor only. Women such as Frida Mgalla hence may still die and be considered to be scanned and diagnosed too late. Yet as a 'pregnancy turned to the back' preserves women's social status as being able to carry children, even if in the future, ultrasound here reduces maternal mortality and morbidity as it grants the survival as a meaningful social being at this particular place.

Contingent children

"A child is blood clots first, then it is a something like a lizard, then it is a child, then in the eighth month it is not a child again, then it is a child again in order to be born a child." Women's recitation about the natural course of pregnancy accounts for children whose primary characteristic, when unborn, is their constant change, not only in size but also and foremost in form. What child is, and whether child is, thus, is by its very nature contingent matter. How does ultrasound cut into this contingency? When I met Nyamwaga Lugari, my question evoked the following dialogue:

> NL: I am happy now. You see, it is a child.
> BMR: It was not a child before?
> NL: It was, it was.... Maybe, you cannot be sure.
> BMR.: You told me that with the kioo you were sure it was a child.
> NL: The kioo showed it was playing, yes it was a child.
> BMR: So after the kioo, you knew you were carrying a child?
> NL: It is a child now
> BMR: So between the kioo and now, was it not a child again?
> NL: A child is a child. You see it here [NL is pointing to the child in her arms]. You remember the picture there, the white round with the dark in the white. You look [at] my baby now. She is white, you see the white [face] with the dark hair. You see now the kioo was speaking the truth.

For women such as Nyamwaga Lugari ultrasound comes in handy, the nurse-midwives at the small rural district hospital in Northwest Tanzania told me. According to international obstetric standards,

Nyamwaga Lugari has a 'bad obstetric history': at age 43, this is her 13th pregnancy; five of her children died either after birth or during pregnancy. Her antenatal record card showed a reproductive trajectory that swayed between home and hospital deliveries. It thereby displayed both adherence to national health care campaigns urging women to deliver their first child and difficult pregnancies in hospital, and adherence to the traditional image depicting a Msukuma woman as good mother only when able to give birth alone, as well as the ability to maneuver between these two cultural demands. After delivering her first child in the hospital, Nyamwaga Lugari gave birth to the following five children at home. When the last two of them died shortly after birth, she attended hospital services once again in her 7th pregnancy again. She then got a caesarean section which made her deliver in hospital with the following three children as well as she considered the operation to be a sign that her womb had gone "tired". Yet, as all these children survived, Nyamwaga Lugari turned to giving birth once again at home. Of the three following pregnancies three children two were "lost in the stomach", as she put it. In order to be prepared for any eventuality, she attended antenatal care in this pregnancy once again: obtaining an antenatal record card for her and many other women signified the passport to be granted access to hospital services should the need arise.

This little dialogue reveals, however, that next to the international obstetric body, another, different body emerged. For me, a midwife trained in 'the West', a pregnancy was of course about carrying a child, and ultrasound simply showing whether this child was healthy and alive. For her, however, the question whether a child was the outcome at all seemed to be in doubt even though the kioo had shown a child as defined by its movements during the examination. During the whole interview Nyamwaga Lugari sat facing me, in her arms the newborn baby girl that she had spontaneously delivered without further ado after having been diagnosed by ultrasound with premature rupture of membranes the day before. For me this image of mother and child had evolved with the ultrasound. Yet, for her the child held in her arms only turned to being the child shown only 'after the fact'. Rather than predicting an outcome, ultrasound for her only retrospectively had spoken the truth. Seen retrospectively her child now was not only that child then but a child at all, and thus my question utterly ridiculous—the contingency of child matter disrupted by the fact in her arms. What Nyamwaga Lugari's ultrasound experience demonstrates is that networks of things and humans may touch or even coincide but they may do so at their own speed—a translation that requires making material sense.

Conceived modernity

In contrast to the stories above, Eunice Hokoro's experience with ultrasound at first glance resembles that of a woman in Europe or the United States. After doing a urine pregnancy test, Eunice Hokoro went to the urban Teaching Hospital to test for pregnancy with ultrasound without waiting for child movements. Her account of the encounter with ultrasound displays the conception of a modernity of young urban professional women in Northwest Tanzania:

> We have open eyes now. We study. We live in town. We are modern women here. You want to know [whether] you carry a child or not? Or are you just feeling different? Maybe your boyfriend wants to know. He wants to know, is this a woman to be married? Can she carry my child? [...] He wants to know is she carrying my child.

It is at second glance that Eunice Hokoro's narrative evokes uneasiness. Rather than ultrasound being a 'couple thing' as in the United States or in Europe, the technology here is used to deliver proof for a spouse to be and to install paternity of a fetus by attributing a technologically objective age to a gestation that matches with the length of a relationship, or of sexual intercourse. Her boyfriend Paul Nabusanke confirmed the power of the images with regard to marriage decision-making:

> A woman might tell you anything. She tells you: "I am pregnant". You marry her. You pay the money. It turns out, she is not pregnant. She has cheated you. [...] So what did you pay for? [...] I want a modern woman, a wife with some education too. But she has studied for a long time, she is not a girl, maybe she has already gone out with other men. So she may cheat on you, telling you that. If you have a test, you know she is truly pregnant. If you have an ultrasound, you know it is a child, it is yours. [...] This is what we do nowadays.

A student of accountancy at the nearby university, Eunice Hokoro was 30 years old when we met. She had aborted two previous pregnancies, as none of the 'fathers' had so far been "the right one", as she put it. Paul Nabusanke, in contrast, "matches": an accountant himself, he would understand that she was an educated, she told me, that is "a modern woman". Her account only points at the concern that this modernity brings with it for her—the implicit doubt that she is able to carry children. This concern is more explicit in Paul Nabusanke's experience with ultrasound: the technology is here seen as part of a modernity that has increased both the prices for women and the difficulty to pay the required bride wealth. Hand in hand with international and individual efforts to provide women with better and longer education, marriage age is postponed—both of which allow for abundant imaginations of a promiscuous life in urban towns, as the two stories illustrate. The fact *that* children are conceived before marriage is not the problem. Yet as marriage becomes more expensive, plights and rights in modern times are now depicted, confirmed or disrupted, by technological means that allow proper insights into bodily matters rather than relying on eyesight. As ultrasound here turns into a maker and/or breaker of marriage, the entity within the body speaks of the social transformation of gender relations in the country. Less than conceiving only a child, or a fetus, Eunice Hokoro's experience with the new technology voices the hopes for conceiving and living modernity.

Eunice Hokoro's ultrasound experience seems to resemble that of women in 'the West': in terms of diagnosing a pregnancy ultrasound enters a network where body signs and calculations (a missed menstruation) are being checked by simple (urinary pregnancy test) to more sophisticated technologies (ultrasound), allowing more insights into what is at stake within her body. Yet ultrasound here also enters a network that consists of cows or money to be paid to a bride's father, of books that might render a body incapable of functioning properly, and of questioned paternity and particular gender relations. The individual body that seems to be subjected to the new technology here at the same time displays a scan of a social body in transition. 'Having a mirror' here is (re)producing body politic (cf. Scheper-Hughes and Lock 1987).

Mirroring pregnancy, embodying society

In many places in Northwest Tanzania ultrasound has caught on, as the fascination of women such as Laurentsia Shitindi and health care professionals demonstrates. Ultrasound, so it seems, has successfully implicated pregnant and professional bodies into diagnostic imaging practices. And yet, some of the bodies that enact or are enacted by ultrasound are quite disconcerting. In order to understand their emergence and persistence within practices to 'mirror' pregnancy in Northwest Tanzania as well as their resistance to a uniformed standard ultrasound body, I suggest, we need to look more closely at the epistemological, ontological and phenomenological questions raised by these new encounters.

"Suppose that there were people who did not invest in the same epistemology that permeates Euro-American science", anthropologist Marilyn Strathern (forthcoming) asks us, concerned with the communication between practitioners and their subjects of care. Following this 'other' epistemology in the encounters of women with ultrasound in Northwest Tanzania, the narratives speak of a very different body that gets voice and form in ultrasound images and people's imaginations. Elsewhere I have shown in more detail how the hospital body that is to meet with ultrasound is enacted in the multiple practices women employ to distinguish between pregnancies eligible for hospital, traditional healer or for home care (cf. Müller-Rockstroh 2007). At the basis of these practices is not the individual body bound by skin. Rather, it is a permeable body, effective to and affected by a range of different agents of afflictions. It is a body that is shared by human and non-human entities, such as ancestral spirits (nzoka), jealousy, worms, as well as by placentas and fetuses. Unborn children may live with them, or may be replaced by them, turning pregnancy-with-child into contingent matter and making the outcome of each new pregnancy a fact only after the actual or the expected delivery. These other entities that in particular pregnant women share their body with, are—perhaps surprisingly—not taken over and replaced by ultrasound and its software claiming to show and measure fetuses only. They are rather informed and supported by the particular ultrasound—the kioo, as Laurentsia Shitindi puts it—even though in the manufacturing hall of ultrasound machines in the Netherlands neither nzoka nor jealousy can be thought of as embodying women's wombs and thus imagined to be depicted by the technology.

In order to understand the continuity of an 'other' epistemology and embodied entities, I suggest, we need to look more closely at the ultrasound ontology. The particular congealment of technological images with societal imaginations described above, I have elsewhere (ibid.) named the 'women's apparatus': this apparatus can "see each and everything" but spares pregnancies to the back from visibility; it sees inside the mother but not inside the child and thus leaves the body of the unborn intact and simultaneously opaque; it may show nzoka and wadudu but only when correctly positioned; it may diagnose a tumor but does so rather accidentally; and it may confirm the existence of an embryo not as a medical but as an economic entity that gains its value in terms of school fees spent and bridewealth gained.

The emergence of the woman's apparatus I have described in the previous section is not accidental: it is the effect of the mutual shaping of the imaging technology with one particular social group, that of pregnant women. At the same time, moreover, it is the effect of previous apparatuses, that is, of previous translations and transformations the technology has undergone while traveling to finally meet with women's pregnant bodies, thereby meeting with other social groups touching and touched by the technology. There is the 'hospital apparatus', for example, which is characterized by particular silences and particular modes of speaking about ultrasound and bodily insides: rather than engaging in show-business, unless teaching of students or of researchers requires it, as in the case of Laurentsia Shitindi, pregnancies in Northwest Tanzania are scanned in silence. This silence is an expression of particular concern for the patient that should be cared for by people who know their history rather than be disturbed with news by ultrasound operators who are, after all—in their and their clinical colleagues understanding—"mere technicians". This silence during scanning the pregnancy is met by speaking about what has been seen afterwards: as nurses thereby aim to facilitate the acceptance of the new technology, hospital practices and national health care programs altogether, local worlds and words are evoked that entangle biomedical with local entities and practices that grant insights into the body. There is also the 'state apparatus' that extends its influence to rural and urban hospitals in order to keep up with international and national decentralization policy, yet fails to systematically embed the new technology into existing practices and programs to improve maternal and child health. And there is the 'international apparatus', promoting a simple diagnostic imaging technology for the complex problem of maternal and child mortality which works even though nzoka and wadudu have not been part of the original set up of the technology. It is by these constant translations that ultrasound work and that particular pregnant bodies are diagnosed and get cared for in new ways. It is thereby the imagination of international experts of the 'poor dying woman in Africa' that sets out ultrasound machines for travel, as it is the imagination of 'white medicine' able of curing all ills that make women come to the hospital. It is the imagination of nurses that women are ignorant and thus need translation, and it is the imagination of women that professionals talk about the same things that indeed people their worlds that makes spirits survive. Pregnant women may quite successfully maneuver between the images and imaginations as patients whose "active passion [...] allows the subject to emerge [...] entangled with and generously gifted by a collective, by objectifs, techniques, constraints" (Gomart and Hennion 1999, 20). While Nyamwaga Lugari follows the trajectory designed for her by biomedical logics (an ultrasound with subsequent c-section), it is only retrospectively that ultrasound is granted the right to have shown truly this child; Janice Mtutu, in turn, uses her new insight to combine the biomedical trajectory with a visit at the mganga, the traditional healer, in order to get rid of nzoka. Women, however, do not blindly follow the reproductive trajectories designed for them by health care professionals. Rather, women act as observant clients and deliberate consumers choosing among the pluralistic health care system. Yet there are also instances where the collective fails and where images and imaginations clash, and end up deadly for mother and children.

Within the practices of imaging and imagining pregnancy, any (pregnant) body in Northwest Tanzania becomes a set of its relations to other entities, a network that consists of conventions, health care professionals, national policies, antenatal record cards, envious neighbors, ancestral spirits, ultrasound machines, stretchers, gel etc. As illustrated in all stories the pregnant body thus is the arena in which particular local socio-technical transformations, conflicting interests, and uncertain relationships condense, and take on form. While the term embodiment has traditionally been used to denote a situation where 'something else' is taking possession of the body, the stories of women in Northwest Tanzania suggest that the term embodiment should be extended to include also an existential condition of social life (Csordas 1994). To be embodied by spirits or by witchcraft concerns perceptual experiences as well as modes of presence and engagement in the world at a particular time and space. Mirroring a pregnancy in Northwest Tanzania evokes and makes visible the embodiment of the technological culture—a particular society.

Concluding observations

Let me reflect about the 'law' underlying the intertwinement of pregnancy and images and imaginations with reference to Foucault's notion of the apparatus as being "a matter of certain manipulation of relation of forces. The apparatus is thus always inscribed in a play of power, but it is always linked to certain coordinates of knowledge, which issue from its but, to an equal degree, condition it." (Foucault and Rabinow 1997, 196). In a context where ultrasound images of fetuses garnish railway stations, enter law suits, and the unborn is made a person with a certain sex and (dis)ability and patients with own rights, a 'researcher's apparatus' may not be able to look beyond the fetus as fact. In the new contexts entered by ultrasound but pervaded by other images and imaginations, the question what ultrasound, and hence, what the pregnant body becomes and embodies is a matter for empirical study. As I have shown this mode of doing ethnography of mediation may lack the security of one locality to which the researcher can go to and become a mere observer. Rather, it calls upon researchers to be aware of the fact that their observing apparatus is one conditioned by the forces that they seek to unravel as 'culture'. To study the body (from) within hence means to study the effect of the intertwinement of existing and new practices to deal with life at hand that turn the body, ultrasound and an ethnography inevitably into situated entities.

CHAPTER NINE

MEDIATED MEMORIES AS AMALGAMATIONS OF MIND, MATTER, AND CULTURE

José van Dijck

Introduction

In the movie *Eternal Sunshine of the Spotless Mind* (2004), the company Lacuna Inc. advertises its method for focused memory removal with the slogan: "Why remember a destructive love affair if you can erase it?" When Joel Barish (Jim Carrey) incidentally finds out that his ex-girl friend Clementine Kruczynski (Kate Winslett) has undergone the Lacuna procedure to wipe their bitterly-ended relationship from her memory, he requests doctor Howard Mierzwiack to perform the same procedure on his brain. Joel is instructed to go home and collect any objects or mementos that have any ties to Clementine ("photos, gifts, CDs you bought together, journal pages"), and bring them to the doctor's office. Upon his return, Lacuna-technician Stan wires Joel's brain to a computerized headset; the doctor holds up each separate object (drawings from his diary, a mug with Clementine's picture, etc.) and tells Joel to let each object trigger spontaneous memories. Stan subsequently tags each object-related memory and punches it into a computer, apparently 'recording' Joel's mental associations on a 'digital map' of Clementine. That same night, Stan and his assistant come to Joel's house, hook up their drugs-induced sleeping client to a machine that looks like a hairdryer but generates images similar to an fMRI, and start the erasure process. As the Lacuna Inc. website explains:

> The procedure works on a reverse timeline, which means it begins with the most recent memories and goes backwards in time. This approach is designed to target the emotional core that every memory builds on. By eradicating the core, Dr. Mierzwiak is able to make the entire memory dissolve.[1]

[1] The movie's producers have created a mock-website of the company Lacuna Inc. at: www.lacunainc.com.

One by one, Joel's memories of Clementine are erased—a fairly automatic process that would have been finished by early morning, if not for Joel's realization, halfway through the procedure, that he wants to keep the good memories of his love affair, so he actively starts to resist the erasure guys. Incapacitated by drugs, he embarks on a dreamlike, psychic journey with a remembered Clementine, creatively hiding her in unconscious, untargeted corners of his memory where she does not belong, in an attempt to escape the high-tech apparatus that is slowly stripping away Joel's recollection of his former girlfriend.

Michel Gondry's fictional treatise of modern science's struggle to erase undesirable autobiographical memories raises important questions: What is the 'matter' personal memories are made of and where are they located? How and why does this matter? *Eternal Sunshine of the Spotless Mind* presents ambiguous answers to these questions, answers that are at once remarkably sophisticated and yet surprisingly simplistic. On the one hand, personal memory is situated inside the brain—the deepest, most intimate physical space of the human body—and the movie actually reflects some state-of-the-art neuroscientific theories on memory formation and retrieval. On the other hand, personal memories seem to be located in the many objects Joel and Clementine (like most of us) create to serve as reminders of lived experiences. Most of these items are what I have elsewhere dubbed 'mediated memory objects', such as pictures, videos, recorded music, diaries, and so on; people have a vested interest in them because they come to serve as material triggers of personal memories (van Dijck 2004). Mediated memory objects, however, are not simply prostheses of the mind, as the movie wants us to believe. Mediated memories, as I will argue in this article, can neither be located within the brain nor outside in culture, but are manifestations of a complex interaction between brain, material objects, and the cultural matrix from which they arise. The body within, in this case the study of memory, will be explored as an intricate amalgamation of mind, matter and culture: mediated memories are concurrently *embodied* through the mind and brain, *enabled* by media technologies, and *embedded* in a cultural context. Rather than leaving the study of autobiographical memory to biologists or psychologists, I will argue that a full understanding of the matter of memory requires the combined perspectives of neurobiologists, cognitive scientists, social and cultural theorists.

Embodied memory: 'Personal memory is in the brain'

Ever since memory entered scientific discussions, it has been caught in the brain/mind dichotomy and appropriated by scholars from various disciplines. Whereas philosophers tend to confine acts of memory to the mind, (neuro) scientists concentrate on the brain as the locus of memory's origin. Until the early twentieth century, the 'location' of memory was generally consigned to the mind, and the 'stuff' that memories were thought to be made of—an indefinable, immaterial set of thoughts and mental productivity—was considered the province of philosophers. From John Sutton's rather impressive historiography of how philosophers from Augustine to Descartes and from Hume to Bergson have conceptualized memory, it transpires that former spatial concepts of thinking about memory have gradually given way to connectionist concepts (Sutton 1998). Metaphors like the library and the archive were commonly used to explain the retention of information or the preservation of experience in an enclosed space, from where it can be retrieved on command (Draaisma 2000). When trying to remember something, the mind, triggered by a material object or image, searches through the stacks from which stored and unchanged information can be retrieved and reread. Research paradigms based on these metaphors assumed memories to be static data from someone's past, and we still often encounter these assumptions in popular representations of memory.

In his important work *Matter and Memory* (1911), the French philosopher Henri Bergson already refuted a one-to-one correspondence between physical stimulus and mental image to account for human consciousness, instead proposing a recursive relationship between material triggers and the images formed by our minds. Bergson's view that memory is not exclusively a cognitive process, but also an action-oriented response of a living subject to stimuli in his or her external environment, prohibits the idea of a 'pure memory' preceding its materialization in a mental image. According to Bergson, "to picture is not to remember," meaning that the present summons to action while the past is essentially powerless; 'recollection images' are never re-livings of past experiences, but they are actions of the contemporary brain through which past sensations are evoked and filtered. In Chapter 3 of *Matter and Memory*, Bergson discusses the relationship between pure memory,

memory image, and perception. In order to analyze memory, he states, we have to follow the movement of memory at work. In that movement, the present dictates memories of the past: memory always has one foot in the present and another one in the future. The brain does not store memories but recreates the past each time it is invoked: "The bodily memory, made up of the sum of the sensori-motor systems organized by habit is a quasi-instantaneous memory to which the true memory of the past serves as a base" (Bergson 1911, 197). In other words, rather than accepting the existence of a reservoir of 'pure memory' from which the subject derives its remembrances, Bergson theorizes that the image invoked is a construction of the present subject. The brain is less a reservoir than a telephone system: its function is to (dis)connect the body, to put the body to action or make it move.

Neuroscientists studying the neurological and genetic workings of memory define the 'matter' of memory by pointing at the brain as the nucleus of all our mental activity and consciousness (Damasio 2003, 183–217). Genes, neurons, and living cells all constitute the bodily apparatus needed to carry out mental functions, for instance cognitive tasks such as factual recall, or affective tasks such as emotions or feelings. But the twentieth century has shown a remarkable parallel shift in understanding the matter of memory between neuroscientists and philosophers of mind. In spite of putting the center for memory activity in the brain, scientists assert there is no such thing as a single location for memory. Even though some parts of the brain are specialized in specific memory tasks—such as the hippocampus for retaining short and long term memory, the amygdala for emotional learning—there is no single vector between one brain system and one type of memory. Autobiographical memory is usually associated with emotional matters that are in turn sheltered by the two amygdalas, yet this does not mean they are solely confined to this part of the brain. Instead, the establishment of memories depends on the working of the entire brain network, consisting in turn of several memory systems, including semantic and episodic memory, declarative or procedural memory (Bourtchouladze 2002). The hunt for the location of memory, undertaken by scientists of various disciplines, has come up with a staggering distributed answer to that question, in fact defying the very possibility of pinning down one type of memory to a single place in the brain. Facilitated by neurological circuits, the brain sets the mind to work, stimulating a perception or a mode of thinking—a mental image, a feeling—that in turn affects our bodily state. The brain is thus the generator of reflexes, responses, drives, emotions and, ultimately, feelings; memory involves both (the perception of) a certain body state and a certain mind state (Damasio 1999).

In more recent philosophies of mind, connectionist metaphors tend to conceive of memory as a distributed agency that leaves traces of an ongoing process. Of all connectionist metaphors that philosophers and neuroscientists have introduced over the years, the networked computer is probably the most prominent one, but not necessarily the best one.[2] Perhaps the symphony orchestra is a more appropriate metaphor than the computer when it comes to explaining the function of memory and how the brain's "matter" is responsible for the personal memories it produces.[3] Like a symphony orchestra needs a brass section, a string section, and a percussion section to perform Mahler's seventh, memory is a function of the *brain* that manifest itself through the *mind* and directs our *consciousness* or conscious acts, such as self-reflection or autobiographical reminiscence; it is a consortium of concerted efforts resulting in a momentary performance. Each member of the orchestra plays his or her part, following the prescribed score as well as the conductor's instructions—their individual performances contributing to the overall sense of harmony. The composer's notational score may be adjusted under the influence of some single parts or as a result of the audience's interpretation or appreciation. Even the 'hardware' of musical instruments may be tweaked to accommodate the performance; material changes in musical instruments inevitably result in subtle performative changes. And, as every music aficionado knows, a symphony's performance changes over time, as each performer tends to interpret the score as well as previous performances through a contemporary ear.

---

[2] The metaphor of 'hardware' to describe the matter of memory may be as tricky as the book retrieval metaphor. It presumes that the brain is a fixed set of neurons and genes that remains unaltered when software is run on its electronically wired system. Yet the living cells a brain is composed of, are constantly changing due to external and internal stimuli. Brain and mind work in tandem to produce mental images, moods, and feelings, and they mutually inform their altered states. Rather than deploying the term hardware, I will resort to the metaphor of 'mindware', a concept introduced by Andy Clark (2001), to counter the potential misconception anchored in the computer metaphor.

[3] The metaphor of the orchestra describing the mind's functions stems from Antonio Damasio (1999, 216). However, whereas Damasio restricts the use of this metaphor to the neuroscientific aspects of the mind, I expand its meaning to include objects external to the body as well as cultural aspects that affect the mind's memory functions.

The extended symphony-metaphor may also account for why memories change each time they are 'performed' by the brain. From recent neuroscientific research we learn that the brain stores emotional memories very differently from unemotional ones. Negative emotional memories are retained in a lot more detail than positive emotional memories. In the case of traumatic memories, they tend to be captured by two separate parts of the brain: the hippocampus, the normal seat of (cognitive) memory, and the amygdala, one of the brain's emotional centers. Hippocampal damage can affect one's capability to form long-term memories, but people suffering from this condition may still be able to recall vague pleasant memories if their amygdala is left intact. Memories effectively get rewritten each time they are activated; instead of recalling a memory that has been "stored" some time ago, the brain is forging it all over again in a new associative context. Every memory, therefore, is a new memory because it is shaped (or reconsolidated) by the changes that have happened to our brain since the memory last occurred to us. Neuroscientists' findings are corroborated by clinical psychologists whose research demonstrates that memories of personal experience are never direct and unalterable copies of past experiences, but are partially 'reconstructed'; self and memory work in tandem to allow us the ability to use our own past as a present resource (Buck and Levine 1998; Levine 1997).

In more than one respect, the movie *Eternal Sunshine* appears in sync with current neuroscientific research, as it demonstrates a nuanced understanding of how the brain forms memories. Scientist Steven Johnson (2004b) states in his review of the film that whereas older movies like *Memento* (2000) still reflect the idea of memory as a kind of information retrieval system, the "emphasis on feeling over data processing puts *Eternal Sunshine* squarely in the mainstream of the brain sciences today." In *Memento*, the main character Leonard suffers from complete amnesia after a major trauma; he fervently tries to reconstruct his past by taking snapshots which he instantly annotates with words—sometimes tattooed on his body—in a desperate attempt to counter the constant loss of information about his own identity and past experiences from his brain. Lost information or memories seem to be fixed in the past, and are fixated in the annotated photographs Leonard keeps producing. By contrast, *Eternal Sunshine* reflects a more complex model of memory and how it is stored in different centers of the brain. Joel's memories partly consist of information that can be erased, yet their emotional core persists. Moreover, his memories are not fixed but morph into new ones. When Joel realizes he needs to stop the erasure procedure, he consciously manipulates the process by taking Clementine to memory-spaces where she does not belong—kidnapping her away from the probing scanner, ushering her into scenes from his childhood that he remembers as being humiliating, painful, or very happy. These intense emotional memories are not so much re-experienced as they are re-written through his recollection. Without the slightest science babble, the movie's assumptions on autobiographical memory are broadly compatible with the reconsolidation theory (Johnson 2004a; Hutchins 1996).

If memory is made of molecular and cellular substance, and is transported through the wired systems of its neurological and sensory apparatus, what, then, is the matter of the mental images produced by the mind that we conjure up when reminiscing? The most basic answer coming from a neurobiologist would be that each mindset derived from the brain is made of the same substances: cells, tissues, organs. Due to the mediation of the brain, the mind and its images are grounded in the body proper. The more sophisticated answer, however, includes a refined description of how the mind and consciousness are functions of the brain. Autobiographical memory involves most parts of that well-woven apparatus and comes in various shapes: the recall of facts (where was I born? What is my age?) is as much part of personal remembrance as is the invocation of a familiar mood or event (do you remember the day your brother was born? How sad her death made me feel?) or the conscious reflection on an earlier stage in life (have I really changed since the age of 18?). In some instances, memory is an affective feeling that accompanies our seeing a picture or a mental picture we have formed in our minds. To the extent that emotions inform our memories, the 'stuff' of memory may be partly derived from the external object itself—a scenic landscape or a picture thereof—and partly from the construction the brain makes of it—the auditory, visual, tactile, or olfactory perceptions in our minds. Neuroscientist Antonio Damasio calls the latter an 'emotionally competent object' referring to the event or object (e.g. *seeing* a painting, a landscape; *hearing* a song) that is at the origin of a brain map and elicits a certain feeling: "this picture makes me feel happy" or "this music makes me sad." Invoking that scene or scenery through one's memory may not change the actual object (the painting, the photograph, the record) but it certainly changes the internalized map of the initial trigger (Damasio 2003, 88–98). Recall and permanent rearrangement of our personal experiences, according to Damasio, play a role in the unfolding of desire. The very desire to recreate an 'original' emotion may be the motivation for changing the brain map: "There is a rich interplay between the object of desire and a wealth of personal memories pertinent to the object—past occasions of desire, past aspirations, and past pleasures, real or imagined." (Damasio 2003, 95)

As neuroscientific research indicates, memory and imagination are not the distant cousins they once seemed: both derive from the same cellular and neurological processes and are intricately intertwined in the 'matter' memories are made of. Memory can be creative in reconstructing the past, just as imagination can be reconstructive in memorizing the present—think only of the many visual tricks people play to perform the cognitive task of factual recall. The function of personal memory, even if restricted to studying its 'mindware', is not simply about recreating an accurate picture of one's past, but is about creating a mental map of one's past through the lens of the present. The contents of memory are configurations of body states represented in somatosensing maps. Living cells producing this mindware are all but indifferent to the processes they condition, and thus, we could conclude, memory is only the trick our mind plays on the brain. As humans, we even tinker with these processes, for instance by inserting chemical substances (drugs) that alter the body's emotional state. Or, as in the science fiction of *Eternal Sunshine*, technicians artificially remove unpleasant memories by deactivating those neurological circuits responsible for undesirable responses conditioned in brain maps. The erasure of the mental image of an experience in Joel's brain activates a desire to thwart the procedure, which in turn causes the 'neurological circuits'—in the brain as well as in the technician's laptop—to go haywire.

Assuming the intrinsic mutability and morphing quality of personal memories, laid out by neuroscientists and tested in experimental clinical settings, I now shift the searchlight of this inquiry to a different aspect of memory's matter. In Damasio's as well as most other neuroscientific theories, the nature and materiality of the *external* object or memory trigger is typically taken for granted. It is obviously not the tangible object they are interested in—the painting, the photograph, the landscape that triggers an emotion or memory—but the contents it represents. Neuroscientists argue the actual pictures become part of the mental maps the brain creates in response to the object, so the materiality of the item does not really matter. But is memory indeed indifferent to the shape and 'matter' of external stimuli and piqued solely by its contents, particularly when it comes to mediated personal memory objects? Is the material artifact that invokes memory irrelevant to mental processes, or does its (changing) materiality have reciprocal effects on the mindware that perceives it? In order to understand personal memory as a complex of physical-mental, material-technological, and socio-cultural forces, we may need to understand its distributed matter beyond its embodied nature.

Enabled memory: "Personal memory is in the mediated object"

Consider for a moment this all too familiar hypothetical question: What objects would you try to rescue from your house if it were on fire? When confronted with this unwanted yet potential situation, many people rate their shoeboxes filled with pictures, diaries, and similar mediated memory objects over, or on par with, valuable jewelry and identity papers.[4] While the latter two are expendable, the first is considered unique and irreplaceable: memory objects apparently carry an intense material preciousness, while their nominal economic value is negligible. The loss of these items is often equated to the loss of identity, of personal history inscribed in treasured shoebox-contents. If you pose the burning house question, asking people whether a mere copy of their original mediated memory objects would suffice, there is a fair chance the answers would be largely negative. Many of us appreciate these items for more than contents only: we treasure the fading colors on yellowed paper, the fumes of tobacco attached to old diaries, the irritating scratches on self-compiled tapes. Apparently, physical appearance—including smell, look, taste, and feel—renders mediated memory objects somehow precious.

Some cultural theorists have located the 'matter' of memory precisely—and often exclusively—in the tangibility of mediated objects. Walter Benjamin (1979), writing on reproducible memorabilia like personal photographs, called them the modern relics of nostalgia, the meaning of which lies hidden in the layers of time affecting their appearance

[4] Don Slater (2001, 82) refers to the results of a market research survey in which 39 per cent of respondents claimed their family photos to be their most treasured possessions.

(Leslie 1999). Some contemporary scholars argue that memory materializes primarily through the technology used to produce mediated objects. Media theorist Belinda Barnet (2003), for instance, prefers technology as the main locus for memory research when she writes:

> There is no lived memory, no originary, internal experience stored somewhere that corresponds to a certain event in our lives. Memory is entirely reconstructed by the machine of memory, by the process of writing; it retreats into a prosthetic experience, and this experience in turn retreats as we try to locate it. But the important point is this: our perception, and our perception of the past, is merely an experience of the technical substrate. (no page number)

Whereas both Benjamin and Barnett acknowledge that memories actually change over time—one in terms of the object getting older, taking on a sheen of authenticity and invoking nostalgia, the other in terms of technology defining and replacing the very experience of memory—they are adamant in restricting their focus on memory to its material and technical strata only. Barnett argues the primacy of technologies in our production and reproduction of memories. Quite a few mediated memory objects require the original technological apparatus upon later recall because that equipment is indispensable for viewing their contents. Think, for instance, about the projector and roll-down screen needed to show your old slides, an 8-track recorder for playing these antique tapes, or, to stay closer to the present, the hardware and software to read the large floppy disks on which you diligently continued writing your diary after buying your first word processor.

Clearly, the inscription and invocation of personal memory is often contingent on technologies and objects, but unlike Barnett, I would locate memory not in the matter of items per se, but rather in their agency, the way they interact with the mind. Paradoxically, the real value of mediated objects and their enabling technologies is often thought to lie in their supposedly static meaning, despite their obvious physical decay; and in their supposed fixity as triggers, despite our constant intervention in their materiality. Memory objects serve as representations of a past or former self, and their robust materiality seems to guarantee a stable anchor of memory retrieval—index to lived experience. But the hypothesis that mediated memory objects remain constant each time we use them as triggers, is equally fallacious as the outdated theory that memories remain unaffected upon retrieval—a theory meticulously refuted by neuroscientists. After all, photo chemicals and ink on paper tend to fade, while home videos lose quality as a result of frequent replay (and even if left unused, their quality deteriorates). In fact, it is exactly this material transformation—its decay or decomposing—that becomes part of a mutating memory: the growing imperfect state of these items connotes a continuity between past and present. Their materiality alters as time passes, but could it be the very combination of material aging and supposed representational inertia that accounts for their growing emotional value?

Besides a sort of natural physical decay, there is a decisive human factor in the modification of (external) memory objects. Like human brains tend to select, reconfigure, and reorder memories upon recall, people also consciously manipulate their memory deposits over time: they destroy pictures, burn their diaries, or simply change the order of pictures in their photo books. Memory deposits are prone to revision as their owners continue to dictate their reinterpretation: a grown up woman ashamed of her teenage scribbles revises details in her diary; a bitter man erases videos of his ex-wife; a grandmother takes apart her carefully composed photo album to divide its pictures among her numerous grandchildren.

The double paradox of a stable yet changing external object triggering a stable yet retouched mental image, appears all too persistent in our cultural imagination. In *Eternal Sunshine*, Joel's and Clementine's desire to destroy their reminders testifies to the human inclination to constantly revise our past. An endearing scene in Dr. Mierzwiack's waiting room, showing tearful clients holding their bags filled with treasured items to be destroyed, signals the intrinsic modifiability of objects as they are constantly prone to manipulation and reinterpretation. People have always used material objects not just to store memories but also to alter them, annihilate them, or reassign meaning to them. Mediated memory objects never stay put for once and for all: on the contrary, the deposits themselves are *agents* in an ongoing process of memory (re)construction, motivated by desire. Memory allows for both preservation and erasure, and media objects can be manipulated to facilitate and substantiate (new) versions of past experience.

The parallel between neuroscientific theories of memory formation and cultural conjectures of memory is far from coincidental; I dare to argue the two processes intersect. For neuroscientists, the mediation of memory happens in the brain where various interacting neuro-sensory apparatuses account for their inherent mutability. Cognitive philosophers add to this theory that memories are mediated not only by the intricate brain/mind orchestration, but by the interaction between the brain with physical, external objects it encounters, including the technologies that help make them manifest. Australian philosopher John Sutton (2002, 138), for instance, defines the locus of memory in the hermeneutics of mind and matter; the biggest challenge in analyzing the cognitive life of memory-things is "to acknowledge the diversity of feedback relations between objects and embodied brain." This view is corroborated by Andy Clark (2001, 141), who argues that memory and its enabling technologies are mutually constitutive; he proposes a cognitive science that includes "body and brains" as well as "props and aids (pens, papers, PCs) in which our biological brains learn, mature, and operate." Both Sutton and Clark regard a mutual shaping of brain/mind and object/technology the inescapable consequence of new neuroscientific insights, and advocate a concerted interdisciplinary research effort to face the challenge of new paradigms created by these findings.

Indeed, I agree with both philosophers that memory is not simply triggered by objects, but happens *through* these objects; brain, mind, technology, and materiality are inextricably intertwined in producing and revising a coherent picture of one's past. However, I think this double-edged concept can still not fully account for the 'matter of memory'. Memories, in my view, are not only embodied by brain/mind and enabled by objects/technology, but they are also mediated by the socio-cultural practices and forms through which they manifest themselves. Although practices and forms are commonly squarely located in the realm of culture, they can impossibly be severed from the other two conceptual pairs. But before we take that layer further apart into its cultural components, let us first look at how media technologies and objects matter as instruments for inscribing personal memory and identity.

Embedded memory: personal memory as part of culture

Mediated memories are material triggers for future recall—produced through media technologies, whether pencil or camera. At the same time and by the same means, however, they are instruments and objects of inscription and communication: devices by which humans seek to establish their own identities in the face of their immediate and larger surroundings. Every historical time frame, as Michel Foucault states, is marked by its idiosyncratic regime of "technologies of truth and self," technologies that

> permit individuals to effect by their own means or with the help of others a certain number of operations on their own bodies and souls, thoughts, conduct, and way of being, so as to transform themselves in order to attain a certain state of happiness, purity, wisdom, perfection, or immortality. (Martin 1988, 16)

Instead of attaching technologies of self to the brain, Foucault argues they are always in and of themselves *cultural*. In Stoic culture, for instance, students wrote letters to friends disclosing and examining their conscientious self in order to establish and test their individual independence with regard to the external world.

Contemporary variants of former epistolary practices, such as e-mails or weblogs, also help construct a sense of self in connection to an outside world. People (and, one could argue, especially young people) wield media technologies to save lived experiences for future recall while shaping their identities in ritualized processes. We take pictures on vacation for later remembrance, but also to convince our friends back home of our relaxed and happily sojourning state; we may want to capture our Thanksgiving dinners on video to document some happy family moments, but a home video concurrently serves to reinforce our notion of belonging to a family. 'Technologies of self' are thus in and of themselves social and cultural tools; they are means of reflection and self-representation as well as of communication.

Foucault's concept may erroneously suggest that media technologies can be regarded apart from their habitual and quotidian use. Naturally, our inclination to take photographs or to write a diary is as much induced by the availability of technologies as by our knowledge of how to use them. As members of a society in a particular historical time frame, individuals deploy a set of practices in common response to their shared social environment and material conditions. Taking pictures, shooting a home movie, or taping recorded music are practices shaped by an internalized cultural logic, unquestioned by one's social surroundings, and performed through seemingly automatic skills. Mediated memory objects provide clues to their social and cultural *function*, thus divulging how people use technologies to produce their own material and representational deposits; these deposits, in turn, 'betray' socio-cultural practices (Winkler 2002). Concretely, a photograph concurrently shows an image and relays information about the habit of taking pictures; a home movie may also reveal something about familial power structures by looking at how various relatives and siblings (often males) take charge of the camera.

Besides signifying socio-cultural practices, memory objects come in shapes which are often mediated by individual invention in response to cultural convention. Letters or family photographs do not arise out of the blue: we write letters because it is an accepted cultural form. Family albums may literally predispose the kind of photographs we take of our children. Looking at a 1868 photograph of our great-grandfather, we may be touched or puzzled by the stern look of a posing figure eyeing the camera. Perhaps more invisible but equally important is to acknowledge this memory object to be the result of a historical practice and form: the habit to have a young adult's picture taken by a professional photographer, resulting in a studio portrait. Cultural frameworks are never stable moulds into which we pour our raw experiences to come out as polished products; they are frames through which we structure our thinking and against which we invent new forms of expression.

The significance of socio-cultural practice and forms for memory formation poignantly surfaces in *Eternal Sunshine of the Spotless Mind*. For instance, Joel is asked to bring pictures, CDs he bought with Clementine, or tapes they made for each other into Dr. Mierzwiak's office in order to be destroyed, because confrontations with these items after his memory's erasure might compromise the procedure's success. Many viewers undoubtedly understand Joel's embarrassment when the technician holds up a mug with Clementine's picture—one of these commodified forms of nostalgia—and empathize with his agony upon seeing him tear up pages filled with sketches and words. When relationships fail, as in the case of Joel and Clementine, the painful division of music collections often appears to be inversely proportional to the pleasure of building up communal preferences. More often than not, the collection and consumption of recorded music is a matter of sharing, and the resulting objects are residues of an intensely social process. Compiling tapes with mutually liked music can be an important part of building up a relationship, just as sharing recorded music with others may be a ticket into peer-group culture. Technologies of self are concomitantly technologies of sharing: they help form bonds across private boundaries, tapping into a communal or collective culture that in turn reshapes personal memory and identity.

To summarize my argument so far, scholars from various disciplines have refuted the truism that memories are images of lived experiences stored in the brain that can be recalled without affecting their content. The cliché of (mediated) objects as immutable deposits triggering fixed memories from a mental reservoir is as outdated as the idea of enduring single memories being stored in particular sections of the brain. Scientists and philosophers agree material environments influence the structure and contents of the mind; objects and technology *inform* memory instead of transmitting it. Memory is neither exclusively located inside the brain, and hence limited to the interior body, but neither can it be 'disembodied', because external bodies and technologies are part of the same mutual affect. To this doubled-edged concept of brain/mind and object/technology I have added a third layer of socio-cultural practices/forms that, in my view, complements the other two. Mediated memories perform acts of remembrance and communication at the crossroads of body, matter, and culture.

Memory as an amalgamation of mind, matter and culture

To say that human memory is a complex problem is an understatement; it is such a daunting, intricate object of research that generations of scholars can hardly be expected to map its mechanisms. Academics from a variety of disciplinary backgrounds, from the biomedical sciences to cultural theory, examine how and why we remember and which basic mechanisms scaffold our processes of recollection. Neurobiologists have delved into the operational physiology of memory functions as they relate to *human nature*. Basic research into genetic, neurological, and cognitive aspects of the brain helps understand the effects of injuries and disease on different types of memory. A small portion of this research targets autobiographical memory and the role of emotion in personal reminiscing. Autobiographical memory has historically been the province of psychologists, who examine its working as an aspect of *human behavior*. Cognitive philosophers and social constructivists point at the importance of *materiality and technology*—particularly media technologies and objects—when addressing the issue of memory. Perspectives magnifying memory as a feature of *human culture* prevail in the humanities, most notably in history and cultural studies. In recent years, cultural theorists have frequently commented upon the role of media in interlinking our past and present, and they have engaged with questions of identity and collective memory.

Most disciplinary perspectives restrict their focus either to the 'body within' or to 'influences affecting the body'. In this article, I have proposed a three-tired approach to memory, a layered model that may help understand the complex, multifaceted and interdisciplinary nature of memory. It is at the nexus of mind, technology, and perceptual and semiotic habits that mediated memories are shaped. In facing such a complex subject as human memory, it is helpful to break apart the components of nature, behavior, materiality, and culture to scrutinize them in isolation. But it is equally useful, at some point, to put them back together again and acknowledge their conjunction. If we want to know what happens "within" the human brain, we cannot ignore what happens 'outside' that body wile definitely affecting it. The question of memory ties together the intricacies of the brain with the dynamics of social behavior and the multi-layered density of material and social culture. Mediated memories means our memories are *embodied* by individual brains and minds, *enabled* by the technologies and material objects that render them manifest, and *embedded* in social practices and cultural forms.

But why do we need such complex a model? What are the advantages of combining diverse and seemingly irreconcilable definitions of memory? Why would academics from various backgrounds want to engage with such multi-layered model before they have answered all or even most of the pressing questions emanating from their disciplinary perspectives? Why would it be advantageous for neurobiologists or cognitive psychologists to turn to cultural theorists or social constructivists (or vice versa) if they want to deepen their insights in how memory works? My answer to these questions is simple: academics can hardly afford to ignore each others' research paradigms in developing new horizons on memory. Since we are facing new challenges posed by complex technical-cognitive-cultural phenomena, such as bioinformatics and genomics, we can hardly rely on old models and metaphors to advance our thinking. The mutual shaping of mind, technology, and culture is an important premise of the proposed model; its modest goal is to pull together relevant findings from these various perspectives to see how they can help redefine immanent multifaceted problems in memory research.

CHAPTER TEN

INTERTWINED IDENTITIES

Gail Weiss

Marking autonomous bodies

The physical boundaries of the human body have historically served as both a theoretical and practical means of distinguishing one person from another, as well as one group of people from another (e.g. social bodies, political bodies, etc.). 'One body, one identity' is a basic legal, ethical, political and social presupposition that we both implicitly and explicitly rely upon in our everyday dealings with others. Conjoined twins and, to a lesser extent, parasitic twins, challenge, or at the very least disturb, this seemingly self-evident truth about human existence. Moreover, as I will argue, the intercorporeal connections that all human beings sustain with one another that are discussed at length by Maurice Merleau-Ponty, are dramatically revealed through the complex ties that bind together conjoined twins. By looking at the various 'paradoxes' that conjoined twins have embodied for their social peers across time and space, we can, I believe, arrive at a better understanding of the challenges intercorporeality poses to traditional, rationalist conceptions of identity.[1]

The very distinction between conjoined twins and parasitic twins and the extraordinary ethical, legal, social, and political dilemmas these people have faced (all centered around the crucial question, "One person or two?"), collectively provide insight into prevalent cultural concepts of personal identity that are so taken for granted, so much a part of the Husserlian 'natural attitude', that they rarely get interrogated as such. Both experiences of exteriority as well as interiority are complicated by a conjoined twin's intimate connection to her sibling's body. I will

[1] This essay is inspired by my interest in bringing Merleau-Ponty's rich notion of intercorporeality to bear on contemporary discussions about identity and marginalization in feminist theory as well as in the rapidly growing field of disability studies.

begin to address the ways in which conjoined twins trouble the dominant understanding of autonomy that underlies popular conceptions of personal identity by turning to French feminist theorist Collette Guillaumin's discussion of marking.

In "Race and Nature: the System of Marks," Guillaumin discusses the multiple ways in which bodies are marked as well as the ways in which the race of white bodies and the sex of male bodies are usually unmarked precisely so that the racial and sexual markings of other bodies can be revealed. Marking the bodies of others, she argues, has historically been intended to designate bodily inferiority, which in turn has been used to justify the social subordination of those whose bodies are marked as other to those whose bodies are unmarked and hence, deemed natural, thereby serving as the standard that all bodies should emulate. Guillaumin distinguishes between voluntary and involuntary markings, that is, those markings that people choose for themselves, such as hair styles, clothing styles, piercings, tattoos, etc. and markings made on individuals' bodies without their consent such as the brandings performed by slave-owners on the foreheads of slaves, and the tattooed numbers Nazis used to identify concentration camp inmates. Guillaumin clearly identifies bodily marking as a strategy used to create social and political hierarchies according to which some bodies are experienced as more powerful than others because of their markings (e.g. royal dress, uniforms, etc.) or the alleged absence of markings (e.g. their white skin, or maleness). However, she does not dwell in detail on the significance of her basic distinction between voluntary and involuntary markings themselves and the corresponding body modifications that they entail.

While cases of involuntary markings appear to be fairly clear-cut to the extent that they are performed without the knowledge and/or consent of the person in question, it is less evident whether any markings can truly be called voluntary if this latter signifies a decision made by an individual agent alone. For, as Guillaumin herself acknowledges, there are always cultural forces at work that help to create the context for specific marking practices; accordingly, individuals' decisions to mark themselves can never be understood outside of the larger habitus (to use Pierre Bourdieu's language) to which they belong and within which their bodily modifications will invariably be interpreted.

All too often, however, contemporary rhetoric concerning 'voluntary' body modifications (e.g. cosmetic surgery, surgical implants, prostheses, etc.), justifies the latter via an uncritical appeal to a Kantian framework that emphasizes the rational autonomy of the person making the decision about what she wants to do with her body. Echoing the classic abortion rights perspective of "it's my body so I can do what I want with it," the emphasis then gets placed on making sure that others don't interfere with my right to mark my body as I see fit and/or on creating a safe cultural space in which an individual can express her difference from other bodies without being denigrated or harmed in any way.

Such a model, I would argue, laudable though its goals may be, also buys into a problematic logic of identity in which my individuality can only be achieved by separating my interests, needs, and desires from those of others. Although the focus of this discussion will be on an example of involuntary marking (e.g. separation surgeries performed on unconsenting conjoined twin infants), precisely because it is here that the damage done by an uncritical allegiance to an anti-relational conception of identity is most readily apparent, one of my goals is to turn our collective attention to the ways in which unexamined commitments to enlightenment ontologies are actively setting the stage for our understanding of both voluntary as well as involuntary body modifications.

Through a critical examination of the case of the Bijani sisters, adult conjoined twins who requested separation surgery and who died in the process, I will also question whether any choice to undergo body modification is purely voluntary. Denying that body modification decisions are solely made as a result of individual choice is not intended to undermine a strong sense of bodily agency, but rather, to challenge whether this experience of agency requires that one's own body be viewed as autonomous from the bodies of others. For my bodily choices never take place in a vacuum; they are always responsive to the situation in which I find myself and, more specifically, they both reflect and affect the ongoing intercorporeal connections that constitute what Heidegger called the *Mitsein*, or my being-with-others in the shared world of our concern.

I will now consider a few specific cases of separation surgeries performed on conjoined twins, in order to reveal how these surgeries indelibly mark not only the bodies of the patients, but also mark a refusal to confront the limits of traditional presuppositions about identity, more specifically, a 'one body, one identity' logic that the very existence of two individuals in a single body calls into question. Since most separation surgeries are performed on conjoined infants (with the notable exception of the Bijani sisters in 2003 whom I will be discussing later), most of these surgeries appear to be clear cases of involuntary marking, at least for the patients themselves who are too young to be consulted.[2] However, unlike the involuntary markings described by Guillaumin, markings which are intended to visibly signify the inferiority of one body in relation to others, separation surgeries are intended to *normalize* bodies, that is, to take bodies that are deemed to be naturally inferior to other bodies and to transform them so that they can be unmarked in their conformity to the nonconjoined bodies of other members of society. The success of this process, of course, depends on the ability of the surgeons and the families to make the massive scars and disabilities produced by the surgeries invisible to others. Even though this often turns out not to be possible (indeed, virtually all twins who survive these surgeries need years of follow-up surgeries and rehabilitative therapy), the idea of both twins, or at least one twin, achieving a 'normal' existence as an *autonomous* individual remains the ideal that motivates surgical and familial decisions.

While there are many conjoined twins who have survived to adulthood and even quite a few alive across the world today, their testimonies about their experiences have not typically been sought by surgeons and the families of conjoined infants contemplating separation surgery. This refusal to gain valuable information directly from those whose lived situation is similar to the patients is not surprising, however, because, as Simone de Beauvoir argued in *The Second Sex* over half a century ago, when societal myths conflict with reality, it is reality that tends to be rejected and the myth that is retained.[3] Feminist philosopher Nancy Tuana has more recently identified this strategy as a part of what Eve Sedgwick originally called an 'epistemology of ignorance', an epistemological framework founded on a refusal of knowledge. Indeed, as Cheryl Chase, Alice Dreger, Ellen Feder and other intersex activists have poignantly revealed, the same strategy has been at work in the case of 'normalization' surgeries performed on intersex infants: parents are routinely denied access to the testimony of other parents

---

[2] I will leave aside the role of the twins' families for the moment though it undoubtedly complicates the issue.

[3] Beauvoir, of course, was referring to what she calls, 'the myth of woman' but I am suggesting that her analysis can be applied to other social myths as well. Here, I am thinking particularly of the myth that identity cannot be achieved without bodily autonomy from others.

who have had intersex infants and who have refused surgery as well as the testimonies of adult intersex individuals themselves.[4]

My own interest in the recent media attention given to separation surgeries performed on conjoined twins in the U.S. and abroad, arises out of a broader, feminist concern with issues of autonomy, identity, embodiment, and relationality that are invoked and troubled by the existence of conjoined twins. Sandra Harding's exhortation to look to specific examples of how 'epistemologies of ignorance' operate in order to come to terms with their systemic influence in our lives is precisely the strategy that I pursue here. (Harding 2004) However, working with the specific sensationalized example of conjoined twins carries with it substantial risks, namely, that of further subjecting these extraordinary bodies to an exoticizing, othering gaze. While I was doing research for this project, poring through books and articles on conjoined twins, I was struck by how often these works unwittingly participate in what Rosemarie Garland Thomson, following David Hevey, terms the 'enfreakment' of these individuals by relying so heavily on freak show handbills and disturbing photographs of various types of conjoined twins to elucidate the contexts in which they have been understood and interpreted. (Thomson 17) Even though the aim of these works is primarily to debunk popular understandings of conjoined twins as freaks and to foreground their basic humanity, the photographs themselves serve to reinscribe the oppressive practices of objectification under which most conjoined twins are condemned to live out their lives. By discussing this danger openly, I am hoping to disrupt what feminist film theorist Gertrud Koch has identified as the 'suturing' of the viewer's gaze to the objectifying, exoticizing gaze of the camera. (Koch 1985) Rather than avoid these images altogether (a strategy doomed to failure since they pervade the media on a regular

[4] Thanks to the work of the Intersex Society of North America (ISNA) in particular, information is increasingly being made available to doctors and to families about the deleterious physical and psychical effects of these surgeries. In fact, some doctors have stopped recommending genital surgery on intersex infants as a result of this information. The ISNA website is an invaluable resource and should be consulted for further details on the surgeries as well as on living life without them: www.isna.org. There are many striking parallels between the case of conjoined twins and the case of intersex infants that must necessarily be taken up in another study. However, many of the claims I am making about the involuntary surgical marking of conjoined infants in the name of a normalization that is never practically achievable for the patient in order to assuage cultural anxieties about identity can also be applied to intersex infants.

basis), I am arguing that mediating our perception of them through critical analysis allows for new ways of seeing and moving beyond entrenched epistemologies of ignorance.

The dominant logic of identity

The 'dominant logic' of 'one body, one identity' creates the context in which separation surgeries are presented as miraculous, altruistic attempts to provide conjoined twins with a particular form of bodily integrity (paradoxically achieved through cutting open and irremediably altering these bodies) that is predicated on the physical separation of one body from another. If one ascribes to this taken-for-granted logic of identity, the refusal of conjoined twins (and more frequently of their parents since most of these surgeries are performed on infants) to opt for separation surgery might seem irrational and perhaps even immoral. It should not be surprising, then, that parents' and doctors' enthusiasm for separation surgery, even when neurological damage and/or damage (or even removal) of other vital organs is an inevitable result for each twin, is typically portrayed in glowing terms. On October 13, 2003, for instance, the Houston News posted the following story on its website, significantly entitled: "Twin Boys Begin New Lives After *Successful* Separation." (my emphasis) The report describes the father's and doctors' reactions to the surgery as follows:

> The boys' father, Ibrahim Mohammed Ibrahim, fainted when he heard the operation was over. "At one point when someone came up and said, 'you have two boys,' the father jumped to my neck and he hugged me and he fainted and I cared for him," said Dr. Nasser Abdel Al, who was with the family for the marathon operation... "He told me that he never dreamt of such a moment," said Abdel Al. He added that Ibrahim's wife, Sabah Abu ed-Wafa, "was crying like everybody else."

A bit later we are told:

> Dr. Kenneth Salyer... said his feelings had ranged "from moments of ecstasy to moments of concern and anxiety." Swift [who we are informed was one of five pediatric neurosurgeons involved in the operation] described an unexpectedly difficult moment separating the left hemisphere of Mohamed's brain from part of Ahmed's. "It was very, very stuck together." He said it was too early to tell what kind of neurological damage the boys might have.

Despite this rather chilling reminder that the ultimate outcome of the surgery cannot yet be ascertained, the report, as its very title bears out, "Twin Boys Begin New Lives After Successful Separation," presents itself as a success story. And yet, the report itself contains the possibility of a counter-reading that enables us to see the surgery as a vehicle to achieve normalization at any price, even if the boys' lives and/or well-being are sacrificed in the process. For instance, at the end of the report we are told that:

> As conjoined twins, Mohamed and Ahmed smiled and giggled, babbled in English and Arabic and tried to move around any way they could. But experts had said they were getting behind in their development, compared with other children their age, because they were unable to explore the world. When doctors concluded that a separation surgery was possible, the risks were explained to the parents. The boys' father told the doctors to go ahead. "If they're left this way, they're not going to be normal," Ibrahim said through a translator earlier this year.

In a separate example from NBC Channel 4 posted on its website on September 11, 2003 regarding two 9 month old identical twin girls who had undergone separation surgery, we see the dominant logic of 'one body, one identity' (and the 'pounds of flesh' it quite literally exacts) again at play:

> "For the girls, it's the beginning of life with two separate bodies able now to participate in society as two individuals," Stein [one of the surgeons] told a press conference afterward. The large intestine could not be divided and went to one twin. Surgeons also performed a liver separation and did reconstruction on the girls' urinary and reproductive tracts. Stein said there were still many challenges ahead for the girls. "Once you separate the kids, the hardest part is how to get them back together. How do you get them in one piece when you are short of tissue?" The next 24–48 hours are critical, but in the coming months and years the girls will see specialists to be fitted for prosthetics to help them lead normal and independent lives.

Perhaps the most well-known contemporary case highlighting the dilemmas of separation surgeries was that of twenty-nine year old Iranian conjoined twins, Ladan and Laleh Bijani, who died in the summer of 2003 in an operation in Singapore to separate them (they were joined at the head). The two women were warned that they might not survive the surgery, and that they might suffer serious brain-damage as a result of the operation. Both sisters were insistent that they wanted to proceed with the operation despite the serious risks involved. Their physical separation, they argued, was worth the possibility that they would not survive the operation. In the end they died, separate in body but as inextricably conjoined in death as they were throughout their lives.

Based on her own substantial research on conjoined twins, historian of anatomy, Alice Dreger, believes that the Bijani sisters are the only adult conjoined twins ever to seek surgical separation.[5] Although some adult twins have, in times of anger, expressed a wish to be separated once and for all from the other twin (for instance, such a wish has been documented on the part of the most famous conjoined twins, Cheng and Eng, and is depicted as the primary desire of Eng in a recent novel about them), no other adult conjoined twins who might be candidates for it, aside from the Bijani twins, have so actively and publicly pursued separation surgery. The majority of these overwhelmingly unsuccessful operations are undergone by infants and small children. In these latter cases, it is the parents and doctors who seek separation surgery, unless, as often happens, the parents abandon the conjoined twins or give them up for adoption (the Bijani sisters were themselves adopted as babies by the doctor who initially advocated for their eventual separation surgery). If the Bijani sisters are the only, or even one of the only, sets of consenting conjoined adult twins to seek this radically experimental surgery, given the fatal consequences of their surgery, we must confront the question of why parents, doctors, and society as a whole deems such surgery to be so desirable even when the overwhelming majority of conjoined twins do not express the extreme dissatisfaction with their situation reported by Ladan and Laleh Bijani. Given that the standard for a successful surgery, as Dreger documents, is merely for the separated twins to survive the operation, and that only 5% of separated conjoined twins are able to leave the hospital after the separation procedure, we must wonder what has raised the level of cultural anxiety so high that the surgery nonetheless seems like the only viable alternative to an unlivable existence, an existence that many individuals nonetheless live, finding love and happiness along the way. According to Dreger,

---

[5] Indeed, in a 1998 article Dreger claims: "I have yet to find an instance in which conjoined twins have sought out separation." (10) Although there may well be other cases of adult conjoined twins who have desired surgical separation, the fact remains that the overwhelming majority of adult conjoined twins have not sought out this surgery.

> The persistent claim behind much of this separation work is that separations must be attempted for the good of the patients—that a life joined is no life worth living (no mind what conjoined twins themselves say). However, I conclude here instead that attempts to separate twins are driven largely by a deep-seated concern for cultural norms of individuality. Bodies whose congenital conformations defy those norms—the bodies of conjoined twins—are treated with surgeries designed to bring the bodies into conformation with cultural norms. (4)

Many conjoined twins, it should be noted, are not candidates for separation surgery at all. Dicephalic twins, conjoined twins who have separate heads but a single lower body, cannot be separated without killing one of the two twins and it should be no surprise that these twins, in particular, have seemed to pose the greatest challenge to the dominant logic of 'one body, one identity'. Even in these cases, however, surgery is sometimes recommended with the ostensible aim of improving the quality of life for the surviving twin by giving her sole use of shared vital organs. These cases of 'twin sacrifice' as Dreger refers to them, involve surgically asphyxiating the twin the doctors determine to be medically less viable and harvesting her organs for her sibling. Strikingly, Dreger notes that "in all of the cases the intentionally sacrificed twin died, but notably, *in not a single case* has the twin chosen to survive ever actually survived to go home or even to live free of a ventilator. Angela Lakeberg [the famous conjoined twin who underwent this procedure in 1993], who was dead by her first birthday, seems to have survived the longest by far." (17) Though there is currently a twin who has survived a sacrifice surgery performed in the UK in 1996, Dreger's observation of how these surgeries have routinely failed should nonetheless lead us to wonder why bringing about the death of one twin when the chances of securing the life of the other is so uncertain is still perceived as such a positive option.

Despite the fact that 1) it is ordinarily illegal to harvest organs from an unconsenting living donor, 2) that most people find such a concept to be morally reprehensible, and 3) that in many cases where the intentional killing of one twin has been carried out, both twins could have lived an indeterminate amount of time without the surgery, it is astounding that doctors are still so eager to perform this type of surgery. The example of twin sacrifice, a procedure that is totally unacceptable to perform on a singleton in order to save the life of another singleton, succeeds in bringing home Dreger's point that corporeal autonomy is so prized as the very hallmark of identity that medical professionals and many laypeople are willing to accept death itself as the price for even one twin to achieve bodily autonomy from the other. And, it must be noted, the surviving twin's total dependency on machines to live even a short time, belies the idea that autonomy is even being achieved through the operation. In Dreger's words,

> In spite of documented cases of reasonably successful joined lives, many singletons, especially surgeons, find it inconceivable that life is worth living as a conjoined twin, inconceivable that one would not be willing to risk all—mobility, reproductive ability, the life of one or both twins—to try for separation. Why, then, is this? (11)

Bodies on display

In *Extraordinary Bodies: Figuring Physical Disability in American Culture and Literature*, Rosemarie Garland Thomson takes up this question, not only with reference to conjoined twins but more generally with regard to all disabled individuals who are typically deemed by the able-bodied majority to be leading lives of terrible misfortune. Rather than take the obvious strategy of demonstrating that the lives of conjoined twins and other individuals with 'extraordinary bodies' are not so bad as they may seem (the project of Daniel P. Mannix's first hand account of living as a freak on the carnival circuit in the second half of the twentieth century, entitled *Freaks: We Who are Not as Others*), Thomson deftly turns our attention away from these 'freaks' themselves and onto the practices of 'enfreakment' that serve to consolidate the position of what she terms the 'normate'. The 'normate', Thomson declares, "is the constructed identity of those who, by way of the bodily configurations and cultural capital they assume, can step into a position of authority and wield the power it grants them." (8) By projecting onto the freak those undesirable qualities that the normate finds intolerable in himself such as feelings of bodily insecurity and vulnerability, and by declaring the unnaturalness and unlivability of this abject subject position, the stage is set for submitting these 'deficient individuals' to a stringent disciplinary regime that is self-justifying (both literally and figurally) for the normate even as it delegitimizes the actual bodily experiences of those whose bodies transgress society's corporeal norms.

While in the past many individuals judged to be corporeally deficient have escaped life-threatening attempts to normalize them because they could not afford the costly surgeries deemed necessary to 'correct' them, the rising popularity of separation surgeries of conjoined twins in order to 'showcase' the brilliance of the doctors and the cutting-edge technologies of a given medical center and even nation, has meant that more and more often these procedures are being performed *pro bono*, so that the patients', their families' or their guardians' economic class need not be an obstacle. However, even the most naïve layperson cannot fail to note through all the media attention paid to the surgeons both before and after the separation procedures, that the surgeries offer incredible career-defining opportunities for the doctors to learn more about the interiority of the body, allowing them to separate that which was indissolubly connected in order to technologically create (like a veritable demi-god) two independently functioning human beings.[6] Thus, unlike most elective and non-elective procedures in the U.S. today (and the U.S. has performed more separation surgeries than any other nation), the hospital, rather than the patients, usually foot the bills. They thereby enjoy not only the worldwide reputation of being able to perform the most complicated, collaborative surgery possible (there are typically over 25 doctors in the operating room alone and the surgeries commonly run longer than 24 hours), but also they receive extra 'brownie points' for providing these life-altering procedures for free to individuals who could never had afforded them on their own.

The material benefits of performing these surgeries then, extend far beyond the interests of the patients themselves, and this cannot be overlooked if we are to understand why they are so frequently performed even when those in the best position to judge if a life conjoined is worth living, are not even consulted to gain their perspective on the issue. As Dreger notes, not only do most adult conjoined twins with the exception of the Bijani sisters view this surgery with horror, but there are also many historically documented cases where, upon the death of one twin, the remaining conjoined twin still refused to be surgically separated from her deceased sibling, even when this procedure had a fairly high likelihood of success and the failure to undergo the procedure signified certain and imminent death. The most famous example of this willingness to die together rather than to live on without one's conjoined twin, is the so-called Biddenden Maids who were born in 1100 and lived for 34 years. "When one died" Dreger tells us, "the survivor refused offers of separation, declaring, 'As we came together

[6] Although, as Dreger amply illustrates, this is merely the goal, not the reality.

we will go together.' This sister then died six hours later." (The Wall Street Journal, 7/9/03)

If we are to take seriously the presence of epistemologies of ignorance hidden in our everyday theories and practices, then it is clear that the unique intercorporeal connections that distinguish the lives of conjoined twins have much to teach us. For not only conjoined twins but, as Merleau-Ponty observes, each of us experiences our interiority through (and not despite) our connections with the bodies of others. If an autonomous body remains the sine qua non without which an individual identity is impossible, then this is only because we fail to listen to those whose bodies defy conventional understandings of autonomy. "The paradoxical fact," Dreger asserts,

> is that *being conjoined is part of conjoined twins' individuality.* If we singletons cannot understand that—if we cannot comprehend a life of two consciousnesses in one continuum of skin—that says something more about us than about them. For we need only to look to history to see that they, too, manage to be human, that they, too, manage to eke out an individualized existence in a very connected world. (26)

The possibility of separating conjoined twins provides a respectable medical medium for resolving concerns about identity and individuation, concerns that are undoubtedly aggravated by the fact that many people in many parts of the world today have a good chance of dying in hospitals hooked up to machines that regulate the very functioning of their internal organs. Political, ethical, and social commitments to bodily autonomy as a mark of identity help to guarantee a view of conjoined twins' existence as irremediably and tragically impoverished. And, if bodily autonomy becomes the very mark of the human, it is surely no surprise that separation surgery appears to be worth any price.

Conjoined twins offer us an opportunity to reassess our own corporeal commitments to specific identity politics. Indeed, they have the potential to reveal especially poignantly a key point of feminist care ethicists, namely, that autonomy is itself a problematic ideal insofar as it forces us to deny the value of the primordial experience we share with conjoined twins, that is, being born connected to another. Adopting a relational approach to individuality, one that emphasizes that this latter can only be achieved through others rather than despite them, is, I would argue, a productive way to combat reductionistic conceptions of 'one body, one identity'. To do this, as Lorraine Code suggests, we need a revision and expansion of the epistemological imaginary, in this case, regarding our traditional understanding of identity as grounded in bodily autonomy. (Code 2004)

The medical and societal emphasis on the urgency of separation surgeries for conjoined twins to realize their individual identities, presumably for the first time, or in more than *name* only since each conjoined twin is given their own name at birth, may indeed turn out to be a primary case of epistemological ignorance, or what Marilyn Frye identifies as "not knowing what we're doing." (Frye 2004) The risk of this type of ignorance is that it reinforces and legitimizes an impoverished conception of our own relational identities as well as the relational identities of the twins themselves. More specifically, conjoined twins incarnate and generate fears of our intercorporeality, the intimate connections between our own identities and the identities of others as they are embodied (that is lived) from moment to moment in our daily lives.

In performing the 'miraculous' technological feat of separating what is depicted as never intended to be joined to begin with, these surgeries enable us to maintain what Kaja Silverman calls a 'dominant fiction', in this case the dominant fiction that we can only be individuated by (forcibly) separating our bodies from the bodies of others. (Silverman 1992) By transforming our epistemological imaginary concerning the possibilities and limits of identity, we can combat the deleterious effects of clinging to the dominant logic of 'one body, one identity'. As we have seen, this is a logic that conflates autonomy with bodily separation from the other, a form of epistemological ignorance that, in the case of conjoined twins, leads to a willingness to sacrifice, or, at the very least, to compromise severely, one or both bodies in the name of normalization and progress.

The active embodied agency consistently expressed both by infant and adult conjoined twins reveals how important it is to disentangle what Lorraine Code calls the 'webs of belief' that support the epistemological imaginary associated with the logic of 'one body, one identity'. Disentangling the dominant logic of 'one body, one identity', paradoxically, may lead us to refuse to privilege the surgical disentangling of the complex intercorporeal connections between the bodies of conjoined twins. Indeed, Merleau-Ponty's famous words which appear toward the end of the *Phenomenology of Perception* take on even more force when applied to the distinctive intercorporeality of conjoined twins for, if we read these words with conjoined twins in mind, we can see even more clearly how the body within can never be separated from the body without:

> True reflection presents me to myself not as idle and inaccessible subjectivity, but as identical with my presence in the world and to others, as I am now realizing it: I am all that I see, I am an intersubjective field, not despite my body and my historical situation, but, on the contrary, by being this body and this situation, and through them, all the rest. (1962, 452)

CHAPTER ELEVEN

FRAMING INTERIORITY: PORTRAITS IN THE AGE OF GENOMICS

Miriam van Rijsingen

> One result of dissolving lines between paragon and parergon, or cynosure and scope, is that we glimpse "truth/beyond mere imagery on the wall".
>
> Joseph Dupras (1996)

The portrait is a popular genre throughout the history of art. It fulfils the needs of those who want to be remembered, and fascinates or accommodates those who want to know or reminisce about the absent other. The portrait gives us visual access to another person promising knowledge not only about his or her appearance, but also about the social and psychological 'being' of the portrayed. Reading or scrutinising the surface intensively, it promises an in-sight, proper knowledge of the person behind the mask, reflecting back some knowledge about the self (Boehm 1985, Bryson 1984).

In the age of genomics artists appropriate new techniques of molecular biology to produce (self)portraits. The promises of these new techniques—true material or genetic information—seem to accommodate the expectation and desire to know the other wholly and intimately, and at the same time gain some proper in-sight into ourselves, as we are all there in the Book of Life. Science seems to allure us with proper (inside) knowledge. Art seems to allure us with the illusion of its reflection, an adornment to the truth. I will show that there is more to this and investigate the borders of art and science, of portraits and 'Life (Itself)'.[1]

I will present three artworks dealing with genomics and biotechnology. These examples *frame* a persistent scientific quest for proper

[1] The term Life (Itself) was coined by Franklin (1993/2003), taken up by Haraway (1997).

knowledge about identity, human Life, or Life Itself.[2] In the first two sections about Mirror, Matter and Script, I will analyse two genomic portraits that make use of different laboratory techniques: in-vitro culturing and sequencing, both in combination with mirroring surfaces. Both works use visualization-techniques from molecular biology to present us with a new form of (self)portraiture, and at the same time delineate and question the scientific techniques and assumptions of proper knowledge (about a person) in a deconstructive way.

In the third section, Growth and Difference, the focus is shifted to the question of 'how you become what you are'. In analysing an artwork that deals with genetic screening, assumptions about genetic information, development and variability are framed. Finally, in the section Framing Interiority and Scientific Truth, I will draw some conclusions about the specific ways in which these artworks present new knowledge about the Self and about Life Itself, not only across the inside-outside divide, but across the disciplinary borders, in the realm of the passe-partout of scientific and artistic propositions.

Matter and mirror: a portrait by Marc Quinn

The first example is the *DNA Portrait of Sir John Sulston* (2001) by English artist Marc Quinn (fig. 11.1). This DNA portrait makes use of a laboratory technique, in which DNA is presented as a material image.[3] The work consists of a small rectangular petri dish with a mirroring frame. Within the dish we see tear-like, wet drops. These are bacterial colonies grown from single cells taken from Sulston's sperm, containing segments of Sulston's DNA on the brink of visibility.

Looking at this work in the National Portrait Gallery in London, Jonathan Jones from *The Guardian* is mesmerized by the visual presence of otherwise invisible human matter and denotes it "a religious icon", "organic jewellery", "a gossamer presence" and "biological

[2] My concept of the frame is loosely based on Derrida's Parergon-work (1978), considering however art as the adornment, or accessory to science as it is often treated in art-science debates.

[3] Rheinberger (1997) stresses the representational difference between in-vitro techniques (image) and sequencing techniques (script) as part of the development of molecular biology towards informatics.

Fig. 11.1. Marc Quinn, *Sir John Edward Sulston*, sample of Sulston's DNA in agar jelly mounted in stainless steel, 127 mm × 85 mm (without frame), 2001, London (NPG6591).

photography".[4] Interestingly, Jones considers the frame in which the matter is presented as a "silver frame" which makes you "feel" that what is presented here is "very significant". In other words, the frame seems a mere ornament (or encasement) to what is made present as a significant object. It represents a boundary. But what is presented is not quite made out: is it biology or is it art?[5] The bottom-line is that this is "a specific trace of someone... fixed permanently". In this case Sir John Sulston, director of the Sanger Institute for genomic research. The critic is perhaps not so much mesmerized by the artwork as such, but more by the way in which science fulfils the "primitive impulse at the heart of portraiture: preservation of a person" (Jones 2001). The website of the National Portrait Gallery informs us that this is an "accurate display of Sulston's *essential identity*" [italics MvR]. What we have here is an artistic proposition, perfectly paired to persistent genetic reductionism.[6]

Jones's commentary invites some questions about framing and the inside/outside divide on two levels. The first level (a) is that of the body itself, as we seem to be presented with a framed material presence, contained signifying matter—the informational building blocks of Life, an invisible inside made (barely) visible. On this level, Jones detects a problematic turn within the portrait: although the represented 'inside' is very significant, we cannot read or make out a specific person. In this visual mode we all look the same.[7] Quinn himself asserts this as he—in line with genetic knowledge—states that such a portrait could be considered as a "portrait of his parents and every ancestor he ever had, back to the beginning of life" (Grunenberg and Pomery 2002). Moreover, humans share more than 99% of their genetic information.

[4] All quotations in this paragraph are from Jones (2001, no pages). With his reference to a religious icon, I assume that he means a relic or reliquary, a material memento kept behind glass and (often) in a silver frame.

[5] Another context should also be considered, that of religion. What is interesting is that both the (life) sciences and the arts have strong religious connotations and recurring debates.

[6] www.npg.co.uk/live/search/portrait.asp?search=ss&sText=Sulston&LinkID=mp59340&rNo=0&role=sit See also Anker and Nelkin (2004, 10).

[7] Although, materially, this is strictly impossible on various levels, as for example the colonies are grown from sperm. But as gallery viewers we are not aware of that fact. The fact that it was grown from sperm though was part of the debate within the gallery about the acceptance of this work. (From a conversation with one of the Trustees, Ludmilla Jordanova, on 14-09-2006) This refers back perhaps to the material quality, the abject status of wet bodily matter.

And still we all look different. So: how particular is the (fragment of a) specific genome to identity? How can we understand the inside/outside divide (or genotype/phenotype) of the body in this molecular age?

The second level (b) is that of art and science. In his commentary Jones mixes biological ("organic") and artistic ("jewellery") references for the material/object presented. It could be both art and biology. The material/object seems quite unframed (or unclear, hybrid) in that respect, although the actual frame of the work belongs firmly to the realm of art (and perhaps religion). Is this then art presenting a scientific object as art? Where does art begin and stops science and vice-versa? In this respect, something should also be said about the commission and the (first) presentation of the work in the gallery.

The work was commissioned and given to the Gallery by the Wellcome Trust, a scientific research Institute that holds strong societal aims. The presentation of the work included an 'ante-chambre' in which a text explained the science involved, accompanied by a running DNA sequence. Going through this *scientific environment*, visitors could enter a small, chapel-like room in which the work was presented in a single ray of light. These were clearly two different sites and discourses—the enlightenment (text) of science and the mystery (image) of art—firmly divided by a velvet black curtain.[8] Yet, standing before the work, this strict differentiating regime was shattered by the work itself. The description Jones gives us is paramount of a mixed, embodied reaction that has perhaps little to do with the scientific framing of the presentation.

I will take up this mixed reaction and focus on the questions of framing by focussing on the significance and function of the actual mirroring frame of the work itself (in relation to what it frames). As we will see, the frame acts only superficially as a container or border. In the mirror we see ourselves reflected (partially), and thus we enter the work as reflections of the symbolic order, our cultural outsides, framing an inside that could be but is not technically ours.[9] I will pursue two levels of understanding the frame: (a) its symbolic function and (b) its material reality.

---

[8] Within the same room as the portrait there were two more traditional representational portraits, two photographs, one of Sulston and one of Quinn, delineating this room even more as the 'art room', possibly satisfying the usual visitor of the NPG.

[9] The way in which the portrait is often re-produced is with a reflection of Sulston in the mirror-frame. In those instances determinism, genetic reductionism, is firmly in place.

(a) The mirror shows us the surface of the body, a body in culture, of symbolic being. It deals with recognition, identity, and the imaginary self. It seems to frame or encase literally material-organic being, delineating a border between the organism and the environment. But this border is also used as a connecting principle by the artist in his other DNA-works, connecting humans (and plants) in a genealogy that is sociologically and culturally conceptualised: for example the horizontal societal arrangement in *Family Portrait (Cloned DNA)* (2002) and the triptych *DNA Garden* (2002) which refers to art historical iconography, altarpieces and to the story of Genesis.[10] Organisms are connected in new ways, as a form of cultural or symbolic technique.

In this respect the works reflects the theoretical propositions of Katherine Hayles and Donna Haraway, who understand the new scientific developments in molecular biology as an 'isomorphism' (Hayles 1999) or 'oscillation' (Haraway 1997) of nature and culture, or natural and cultural techniques. The in-vitro technique itself underscores this and should be understood, according to Rheinberger (1997, 271), as a *representational technique* in which in-vivo (inside) processes are not mirrored or depicted but "realised", "made available" outside the body. Here the inside is *produced* as outside, the invisible produced *as* visible, in order to produce epistemic objects. The in-vitro cell is therefore not natural, but artificial, it is an "*experimental controlled system*" (Rheinberger 1997, 274).[11]

Using the mirroring frame, Quinn asserts the artificiality and conceptuality of what is presented as significant matter. It allows for and alerts us to changing genealogies. Therefore and at the same time the frame is not a simple symbolic border—reflects not a conventional outside—but is itself a site of production where *something*, cultural or genealogical possibilities, is produced in accordance, or analogy even, with the matter in the petri-dishes.

(b) The material reality of the mirroring frame underscores this and is as important as the material reality of the presented genetic matter

[10] Family Portrait shows four petri-dishes with the DNA of Quinn, his girlfriend, their son and her daughter. DNA Garden shows 77 petri-dishes containing DNA of 75 plants and 2 human beings. For images see Grunenberg and Pomery 2002.

[11] Rheinberger (1997, 2005) investigates the representational dynamic in the laboratory. With Latour, Hacking and Goodman he understands representation as a phenomenon of transport. The image is not duplication, but formation of specific synapses allowing presentation, a continuous process of realization. For theorizing the representational system he uses Pierce's semiotic model of symbol/icon/index.

according to the artist. Polished metal is important in Quinn's work and it has two characteristics. Scientifically, the reflecting surface of metal should be understood as a moving sea of electrons, easily associated with what our body is essentially: water/fluid.[12] Furthermore, it shares with DNA the quality of indifference: "DNA neither cares, nor knows, it just is", as Quinn paraphrases Richard Dawkins (Grunenberg and Pomery 2002). The indifference is abolished only in the coincidence of the body before the mirror.[13] This makes a firm specific symbolic reference or context almost impossible, and is at the same time a critique of genetic reductionism. Again, the frame is not as stable a border as we expected. And it is not as *determined* as we might have assumed.

Exploring the boundaries of material (in)visibility and symbolic reference or frame, one comes up with a oscillation between object and frame. Matter and mirror both belong to representational systems in which something is produced, that refers more to a continuous chain of significations, realisations, becomings, than to particular being. The 'real stuff' in the petri-dish can be of John Sulston, but is nevertheless a *representation*, that is scientifically *produced*, a fragment and trace pointing not to one, but to many (possible) referents in time and space if we understand Rheinberger (2005, 23ff) correct. Both frame and genetic matter are productive themselves and defer images as (mere) duplications, for they are (in)formational realizations. They make something available for interpretation (Rheinberger 1997, 269).

As an artwork, Quinn's portrait of Sulston is not just presenting a single person, the 'real' Sulston. It is far from such reductionism. Instead it explores the borders between I and Other, inside and outside, art and science, just to find they have no definite locality. Or, as Johnson puts it, "The frame... becomes not the borderline between the inside and the outside, but precisely what subverts the applicability of the inside/outside polarity to the act of interpretation" (Johnson 1980, 128). I will come back to this in the last section.

[12] Compare for example The Etymology of Morphology (1996). This work though, is not made from metal, but blown glass, using breath as the 'inner principle'. Much in Quinn's oeuvre is about the changing, metamorphosing body, about the constant oscillation between 'natural' and 'artificial' formative processes.

[13] See for example his work Mirror Self Portrait (2001–2002).

Script and mirror: a self-portrait by Andreas Horlitz

The second example is an *Autoportrait* (1998) by German artist Andreas Horlitz (fig. 11.2). It is an engraving on partly mirrored glass, mounted to the wall in a way that leaves some space between glass and wall. Those viewers who are interested in genetic science will recognize from popular publications what it is: a genomic sequence. More precisely the work contains three extracts from DNA sequences, which are represented in a horizontal montage. The sequences (from left to right) are taken from myosin (responsible for the activity of the heart muscles), keratin (present in the skin and hair), and the analgesic beta-endorphin hormone (formed within the brain). The extracts are such small parts of the sequences, that they are more or less identical to everyone's data (Netta 2005, 112). Not many people will be able to read the work scientifically, but every viewer will see him—or herself reflected in de little mirrors that represent the code. It is important to note that in this case we do not have a mirror as frame for genetic matter, but genetic script as mirror. Before entering on an interpretation, we should look more closely to the scientific and artistic techniques that are used.

Sequencing is a chemical/photo-graphic technique that allows us to *read* genetic information as the (material) text that informs the growth of the organism.[14] It belongs to a particular development in molecular biology that is fundamentally informational (instead of chemical)—hence the text or *script*. It made gene-technology and recombinant DNA technologies—or wet technologies—possible.[15] Script became more than ever fundamentally related to material life itself, because living systems can be identified as structures that process information.

---

[14] For sequencing the building stones of life (DNA) are broken into pieces through enzymatic manipulation, in which radioactive tracers are used. These pieces run into a porous, polymerised plate, are then transferred to a filmic plate, on which the markings of the radioactive tracers appear as black stripes to be identified separately as G(uanine), A(denine), C(ytosine), and T(hymine). The chemical transformation that is used to make something (graphically) visual on a two-dimensional surface, reminds us of photographic techniques.

[15] Rheinberger (1997) draws a demarcation-line between early (or traditional) molecular biology (until the 1960s) and recent forms of gene-technology. The first is fundamentally chemical and the second fundamentally informational. In the second the organism itself becomes a laboratory, a newly defined representational space, as information loaded macro-molecules are inserted in and working within the cell itself. The result is that we are dealing with intra-cellular representations of an extra-cellular project: the outside is produced as inside. This brings about a total collapse of nature and culture.

Fig. 11.2. Andreas Horlitz, *Autoprotrait DNA*, engraved mirrored glass between stainless steel rods, 300 × 220 cm, 1998.

It is not strange therefore that many refer to this text metaphorically as a *set of instructions*, a *blueprint*, or *the book of life*.[16]

The sequencing technique underscores the idea that we could read and understand the life of this specific person—past, present and future. The aim of the Human Genome Project (HGP) when it started in the 1990s is to be considered in that line of thought: to understand human life ultimately as a (material) text to be decoded. An individual embodied life is made from the inside out in this deterministic model of gene action: "Identity is formed and can be expressed in one direction only—from the gene(s), which *are* or *contain* the necessary program of information, to the individual embodiment" (Scully 2006, 352).[17] We should understand it as a form of inscription. But this model provides also a single site: a one-to-one mapping between genes and identity, in which the inside-outside divide seems to play no role (Scully 2006, 353).

However, both interpretations of the deterministic model require an essence to be contained within the genome. The persistence of this model is very strong due to the metaphor of text-reading, although a re-conceptualisation of genetic theory, *a re-reading of the molecular paradigm*, is imminent according to Neumann-Held and Rehmann-Sutter (2006). The reading of the text of life proves not to be so easy as assumed. If we understand living systems as structures that *process information*, we should understand them as structures that (re)read, (re)interpret and (re)write *continuously*. "Being is writing" and therefore "there is no script without transcription", according to Rheinberger (2005, 17, 18).[18]

Before returning to Horlitz, something more should be explained about sequencing as a representational technique. Because we are bound to believe that science produces material facts and art produces metaphorical images, but with sequencing those particular paradigms are not exactly working. Moreover, the text we are referring to is not a text as we know it. In sequencing something is made visible, more precisely, it

[16] See for example Kevles and Hood (1990); for a historical analysis see Kay (2000); The metaphor is very persistent: genes can "spell out a recipe...for building, growing and running a living human body", brochure of the Centre for Life, Newcastle (1999), cited in Scully 2006, 352.

[17] Scully (2006, 352) calls this the Watson-Crick central dogma: DNA makes RNA makes protein makes person.

[18] Translated from German by MvR: "Unsere Seinsmachine (ist) eine Schreibmachine" and "Es gibt keine Schrift ohne Überschreibung".

> *produces* a piece of DNA *in the form of* a text, *realizes* it as a sequence of letters, but not on the well known level of the alphabet or the printed page or a language, but as genetic script of which the primary traces are being made visible. (Rheinberger 1997, 278, italics MvR)[19]

The text is made visible by tracers, it is itself a trace, but with no stable referent. It is rather more productive (as the living organism itself is).

> The biologist/researcher does not work with the genes as such, but with produced 'graphemen' in a representational space. If he wants to know what they mean, he has no other possibility than to interpret the articulation of these 'graphemen' with others. (Rheinberger 2005, 22–23)[20]

Sequencing is therefore also (representational) *writing*. We should read the textual sequence, as the organism itself 'reads' the genetic text. And here is where determinism is hard to follow. Reading is (in the organism as well as in its laboratorial representations) always and continuously a process of *translation and transcription*—instead of inscription. [21] It is a continuous process of reading/interpreting/writing in which the very idea of an essence is constantly frustrated or suspended (Rheinberger 1997, 269–276).[22] It is a continuous process of constructing and deconstructing. I will specifically come back to this issue in the next section of this text.

Returning to the *Autoportrait*, we are confronted by a (graphically abstracted) text of which we can all too easily assume that it gives us Andreas Horlitz essentially (Netta 2005, 106). We seem certainly invited to understand the engraving technique as a form (and analogy) of

---

[19] Rheinberger refers to the 'non-metaphorical' character of this script In his later work (2005) he differentiates between several historical 'textual systems' (Schriftsysteme); genetic script is not understood as imaginative/referential duplication nor as notational/linear 'language'. Important characteristic is that it is independent from imprint or expression, seeing or talking. Rheinberger uses the Derridian term "Graphemen".

[20] Translated from German by MvR: "Der Biologe, als Forscher, arbeitet nicht mit den Genen der Zelle als solchen, er arbeitet mit experimentell in einem Repräsentationsraum produzierte graphemen. Wenn er wissen will, was sie bedeuten, hat er keine andere Moglichkeit, als diese Artikulation von graphemen durch eine andere zu interpretieren."

[21] This is specifically worked out in biosemiotics. The cell is looked upon as an interpreter of DNA. Cellular interpretation refers to "the formation of an interpretant in the form of the specific activity of that complex of finely regulated and membrane-associated enzymatic reactions which collectively are known as transcription and translation" (Hoffmeyer 2006,156).

[22] Rheinberger points specifically at gene-technology as a wet technology, in which information-loaded molecules enter the organism in order to read/interpret/(re)write the genetic text. The slightest idea of essence, as well as the idea of nature, is of course frustrated by this: inside-outside, as well as nature-culture are made ambiguous here.

(genetic) inscription on a glass body, to frame this person transparently as its text. But the large glass body itself is only semi-transparent and literally unframed, it is fragmented (in three pieces), and reading the text as indexes for specific referents, the heart and the brain envelope (squeeze) the skin (between them).

And on closer inspection, there's more. It is not the genetic text that is etched on the mirrored glass. There was no inscriptive engraving. Actually the metallic mirroring surface of the glass was etched (sand-blasted) away between the strips of text until the mirrored glass (body) became opaque.[23] The text is presented in the form of little mirrors in which we see ourselves (and the space around us) reflected, forcing us not only to understand this text as a form of (our own) selfportraits, linking self and other as is suggested by Netta (2005:110), pointing at the shared genetic make-up of all humans, but challenging us also on a more profound level of understanding the representational modes of science and culture. Reflecting on the idea of embodiment through (scientific) script and/as (cultural) image, and presenting it as productive space.[24]

Because genetic script is presented and experienced as (reflected) image of the body/space before the work, it invokes a strong *embodied experience* through the space it occupies ánd reflects. The influence of light and movement are crucial to the understanding of the work. When the light changes, the glass screen as spatial object changes with it: the shadows of the mirroring text moves across the wall/space. Also, when you move your body in front of the glass, the spatial effects change. Horlitz *Autoportrait* evokes the sense of a semi-transparent or opaque moving and changing body, never ultimately pinned down by the script. Moreover, the text (itself) 'gives space' to that moving and changing body. Script is understood as continuous becoming, an act of (re)reading/(re)interpreting/(re)writing. Here the same applies as in

---

[23] 'With a photographic technique the DNA texts were applied to the reflecting faces of glass mirrors: the mirrors were covered with a photosensitive coating and exposed to light. After subsequent development, the spaces between the strips of the code were exposed, and the metallized surface could be removed in these areas by sandblasting. Here the glass has an obscured translucent surface, not transparent, but allows light to pass through' (Netta 2005, 108).

[24] In the book on Horlitz much emphasis is laid on the suggestion that his works are located "somewhere between representation and abstraction, between concrete information and the world of mental concepts, and that they function on various plains of perception".

Quinn's work: the mirrors are both indifferent and particular, reflecting possibilities of being. Every sense of essence or deterministic containment is lost in the embodied space of Horlitz's work. The genetic text literally takes the environment in, suggesting that the informational in genetics is never disembodied.

This seems paramount for Horlitz's oeuvre; he always reflects on the (transparent) layering and shifting of various sign-systems (texts, codes, languages, visual iconography) in connection with embodied experience, space and time. For example *Text DNA* (1997), *Cyclus* (2000), *Matrix* (2001), *Panorama* (2001) and *Genesis* (2003) are all focussed on graphical scripting or measurement techniques (from genetics, neurology, chrono-biology, but also astronomy and geology), using light and mirrors, running through spaces, marking and shifting space and time.[25] The work that reflects Horlitz's methodology best is perhaps *Palimpsest* (2002), in which no essential or ultimate referent is to be found, instead (re)writings are paired to (re)interpretations evoking (re)readings on a purely visual and embodied level.

Both Quinn's portrait and Horlitz's selfportrait point at deterministic or reductionist tendencies in genetic research, and could be interpreted superficially in that way. Both, however, undermine at the same time that tendency by stressing the processes of interpretation involved in representational techniques and the production of knowledge. They both use the mirror as a cultural technique to deconstruct determinism.

Growth and difference according to Amirthi Perera

Not every artwork that deals with genomics or molecular biology uses or presents scientific artefacts or techniques from the laboratory. *Screen* (2001, fig. 11.3) by Amirthi Perera, is a low tech work, but nevertheless addresses a problem right at the core of genetics, questioning knowledge production ('determinism') in genetic screening. Basically: What can we know about the development of an organism?[26]

[25] For images of these works see Horlitz (2005) or the website of the artist, www.andreashorlitz.de.

[26] The work was shown in the TwoTen Gallery of The Wellcome Trust in an exhibition titled Working Drafts. Envisioning the human genome. The Gallery shows artistic and scientific representations in mixed setup.

Fig. 11.3. Amirthi Perera, *Screen*, 81 ceramic pods 6 cm each flat on a white surface, 2001.

The work consists of 81 sculptural objects. These are glazed ceramic casts, arranged in a square with regular intervals on a low platform. One of the casts is set aside, away from the others, but within easy reach. Viewers are invited to make an alternative choice as to which piece should be set aside based on an interpretation or valuation of difference. Looking at the casts we are reminded of seeds, or even small foetuses. The title of the work, Screen, is in this case clearly a verb, an imperative from screening. It refers to genetic screening, specifically prenatal screening. But also the second meaning of the word, as in digital screen for example, is not totally lost.

Both meanings confront us with problems attached to genetic screening: it confronts us with a reading problem—how to interpret variation? Furthermore, how to interpret the variation of the surface in relation to what it means in terms of informational reference or essence? That is, what can it tell us about (future) development? The most important aspect of this work, then, is its performative quality. It invites the viewer to act upon it, to value and to choose. But what are our motives for this particular choice? Are we trying to read and diagnose ugliness, disease or abnormality, or is it beauty, health and normality? Finally, we must admit to the absence of a clear referent or context. We are stuck with variation and mutation, with pattern and randomness.

Although the work is low tech, something should be said about the production process, because it mimics in a sense—artificially—what it is about: the growing organism. These are all identical ceramic casts, replica's of the same basic material form. But they all have a different outlook, have different *expressions* (a word also paramount in genetics), due to chemical reactions in the process of glazing and heating, but also from unintended draughts in the oven. With the same basic materials random differences appear in the course of the production process. With its dimensions variable, the work can grow infinitely.

To understand the work we should look at the practice of genetic screening and to the more profound questions of the growing and developing organism. We will find that there is a strange coincidence of relativism and determinism. Genetic screening is a practice in which genes are scrutinized. Defects and mutations that occur are interpreted. This is a reading practice with many fuzzy areas. Basically science is still unsure about the exact relationship between genetic information and material expression. When a mutation is found, this does not necessarily mean that the person will develop a/this disease. As research, genetic screening is done precisely to uncover the genetic heterogeneity of humans as it relates to its physical variation. Although many mutations are injurious, there is no ideal referent, no 'genome of genomes', no perfect or *normal* genome—we are all mutants (Leroi 2003). Leroi prefers to talk about variations or polymorphism.[27]

Growth can be characterized by pattern, in which randomness occurs. Something can happen, but we don't always know why. The process can be influenced by both biology and environment. According to Coen (1999,16), the process of the organism can be analysed as a mechanism of copying and creating.[28] Which can be compared to Rheinbergers process of (re)reading/(re)interpreting/ (re)writing. Both Leroi and Coen insist that polymorphism is a fundamental rule (but not a fundamental relativism). All mutations reveal or disclose something of the functioning of the genes that they influence. Therefore, the growing organism itself can be analysed as a deconstructive mechanism of the body.

Going back to Perera's work though, we are not just confronted with variation or polymorphism. The artist herself says, pointing at social and ethical aspects,

> I am interested in exploring attitudes towards the concept of 'abnormality', specifically with reference to the identification of 'disease genes'. Where should the line be drawn in the quest for genetic perfection? And what are the implications of this research fort the unborn child and people who are transmitters of hereditary disease? (Working Drafts 2002, 8:20).

We have to make a choice, and therefore leave fundamental research behind and enter the applied form of medical diagnosis. Genetic screening is foremost a form of diagnostics and statistics to detect diseases, to differentiate between what is normal and what is not. It is all focussed on identification, interpretation and probability, based on collecting as much data as possible—about family history for example, to elucidate patterns of transmission. The referent in the diagnosis is a growing and ever changing statistical (institutional) standard (Duden and Samerski

---

[27] His calculus is astonishing for a lay public: every newborn houses some 100 mutations average, which does not come from one of the parents. 4 of them could change the expression of the genes, of which 3 could be injurious. Added up with the mutations in the parents and their parents the total burden is enormous (Leroi 2003, 80).

[28] Coen (1999) compares the processes of the organism to the processes of art-production. He also shows that cloning, as a form of copying, will never produce an exact 'same'. It is only in the public narratives that the timeless monster of the clone is bred.

1998, 83).[29] We carry specific genes and mutations of genes, but the meaning of them (in screening) is just a cipher in a bundle of statistics, a calculus about the healthy body. The point is, that in genetic screening, *determinism is both constructed and deconstructed.*

Returning to *Screen*, in the invitation to act or play, we are confronted with a chain of three fundamental (related) problems of genetic science: (a) there is no ultimate referent for variability, (b) we should rather consider pattern and randomness than presence or absence,[30] (c) how do we asses the "play of possibilities"—that which is possible x that which withdraws itself from control (Rheinberger 2005, 26–59).[31] Our diagnosis and intervention in a way is a re-enactment of reading, interpreting and writing. The objects on the low stage are meaningful, but floating signifiers, surfaces where the organism meets the environment, and in that sense also interfaces.

But we should not understand this surface as the surface of the closed skin, containing our insides, our code. With Hoffmeyer (2006, 156) we could follow the suggestion that as the inner locus for life we should take up the concept of "the extended membrane". "DNA", Hoffmeyer says, "cannot know where in the organism it is located. Such 'knowledge' has to be furnished through the communicative surfaces of the cells" (Hoffmeyer 2006, 165). Life unfolds across membranes, "and in that sense even the inside of a membrane in a way is also the outside" (Hoffmeyer 2006, 168).[32] In other words, environment is everywhere, and knowledge is produced across membranes constantly in interpretative acts.

[29] In their book, Duden and Samerski question the loss of embodied experience in those new diagnostics and the way in which genetic counseling often develops into explaining statistics and increased risk to startled patients/parents. Franklin (1993/2001) specifically stresses that knowledge and possession of genetic information represents power, opportunity and responsibility.

[30] Katherine Hayles (1999:29–30) makes a paradigmatic philosophical case of this difference.

[31] Rheinberger (2005: 26) specifically draws on Francois Jacob and Derrida's Grammatology.

[32] "If we move inside the surface of a multicellular organism we run into new surfaces, enveloping tissues, organs or single cells. Here we meet virtual worlds constructed by the macromolecular machinery of the cells in order to cope with cellular environments. The cell membranes act as interfaces through which signs from outside are converted into cascading processes on the inside... another step inward, through the cell-membrane, we immediately run into new membranes belonging to cellular organelles like mitochondria, chiroplasts, lysozomes etc." (Hoffmeyer 2006, 164)

Framing interiority and scientific truth

In this article two cases were made. One was that of the artistic portrait and the concept of individuality. I suggested that a portrait tends to give us some insight into a specific person. The idea was that new representational or visualizing techniques developed in science, suggesting to reveal the actual or material essence of (individual) life, would be appropriated by artists. A literal visualization of essential humanity. But I hope I made it clear with the works of Quinn, Horlitz and Perera, that it is much more complicated than that both in art and in the science involved. It brought on the second case, that is the way in which art and science relate and produce knowledge.

The second case is important because the art-science debate tends to formulate either a solid divide between the two, in which art can only be illustrative of scientific facts (or results), a mere ornamental duplicative mirror awed by the science, translating the scientific message to a wider audience; or a hoped for interaction in which differences are played down, and knowledge production takes a new turn, known as the third way.

The first proposition in the debate is most strong in popular discourse and I wanted to focus on that with the question how art and science signify and "where truth is". This proposition assigns science as the speaker of truth in contact with the inner core and significance of fact, and art as an ornament or supplement to that truth. With the idea of the *frame* I tried to show that those two positions (science as manufacturer of truthful facts and art as reflecting those facts) are not as stable, not as delineated, as we think they are.

Take for example Quinn's portrait of John Sulston. The *truth of the matter* is displaced in many ways, in the work itself and in its presentation. The mirroring frame did not quite keep the stuff at its place, because although this stuff can be of no one but Sulston, in this particular representational mode we all look the same, while the mirror (frame) shows nothing but variation and particularity.

In Horlitz's portrait the scientific representational mode that is used is deconstructed in the act of looking and evaluating: it represents Andreas Horlitz, but only in a fragmented way, and only when we read the code as a proper sequence, which it is and at the same time is not. Reading the little mirrors as a code will get us nowhere, but looking 'into' it and understanding them as visual, spatial and embodied reflections will. The work *is* not a scientific text and it *does* something different—or doesn't it? It sets us thinking.

Something should be said about mirroring. Rheinberger has pointed out that in the new developments of molecular biology scientific representation is about taking and making, realizing, not mirroring. Artistic representation on the other hand seems to be about mirroring. Having defined that difference, analogies tend to show themselves, certainly in the works I presented here. The mirror in artistic representation takes on characteristics of scientific representational modes (it is productive as well), and at the same time remind us of our embodiment, from our smallest particle to our being in the world.

Reminding us of our embodiment is part of every work I discussed here, also the last one by Perera, in which the performativity of the work, the challenge to play with possibilities, gives us material to think about the changing, transforming organism that our body is and what is means in the world.

These artworks are part of the art-science debate itself, they propose possible readings in which we are forced to change position between what is presented and what is reflected, *to interpret*, and become "edgy readers" as Dupras (1996) calls it, readers who see themselves suddenly reverse positions between what is (re)presented and what (re)presents it. Truth is displaced in many ways and art and science are neither solidly divided nor harmoniously the same.

The artist Helen Chadwick narrates herself across that border in 'Soliloquy to flesh'.

> On impulse, I plan an incidence of self, in other words, a building site to develop in. As I proceed, things appear to change. A site curves around my presence whilst I in turn mould the geography of space. In mutual circumnavigation, the terrain waxes open, and following the path of least resistance, performs new convolutions. I mirror these curvatures. The architecture grows corporeal and I am enfleshed. In stories of origin and memory, cycles of transformation and decay, I narrate myself as envelopes of feelings quantised into flesh. There at the heart of these inscapes are enucleate abstracts eluding definition, stubbornly refusing to cohere... In parabolas, complementary functions of inside-outside, I plot the amplitude of difference, criss-crossing through permeable screen... Inside is outside is inside. (Helen Chadwick 1989, 109)

BIBLIOGRAPHY

Andreas Horlitz. Arbeiten Works. 2005. With essays from K. Honnef, I. Netta, S. Omlin, H. Scheurer and S. Trescher. Nürnberg: Verlag für Moderne Kunst.

Andry, Nicholas. 1700. *De la Generation des Vers dans le Corps de l'Homme. De la Nature et des Espèces de cette Maladie, les Moyens de s'en Preserver et de la Guerir*. Paris.

Anker, Suzanne, and Dorothy Nelkin. 2004. *The molecular gaze. Art in the genetic age*. Cold Spring Harbor & New York: Cold Spring Harbor Laboratory Press.

Appadurai, Arjun. 1986. *The social life of things. Commodities in cultural perspective*. Cambridge: Cambridge University Press.

Aristotle. 1968. *On the soul*. Oxford: Clarendon Press.

Augé, Claude. 1897. *Nouveau Larousse illustré*. Paris: Librairie Larousse.

Bakhtin, Mikhail. 1984. *Rabelais and his world*. Trans. Helene Iswolsky. Bloomington: Indiana University Press.

Bargheer, Ernst. 1931. *Eingeweide: Leebens- und Seelenkräfte des Leibesinneren im deutschen Glauben und Brauch*. Berlin: de Gruyter.

Barnet, B. 2003. "The Erasure of Technology in Cultural Critique" *Fibreculture* 1: no page numbers. See http://journal.fibreculture.org/

Barthes, Roland. 1981. *Camera Lucida: Reflections on photography*. New York: Farrar Straus and Giroux.

Bartholin, Thomas. 1673. *Acta medica et philosophica Hafnensia*. Copenhagen.

Baszanger, Isabelle. 1998. *Inventing pain medicine: from the laboratory to the clinic*. New Jersey: Rutgers University Press.

Beauvoir, Simone de. 1989. *The second sex*. Trans. H.M. Parshley. New York: Vintage Books.

Benjamin, W. 1997. *One-way street and other writings*. London: Verso.

Benthien, Claudia. 2002. *Skin: On the cultural border between self and the world*. Trans. Thomas Dunlap. New York: Columbia University Press.

Bergson, H. 1911. *Matter and memory*. London: George Allen and Unwin. (Originally published 1896).

Bidloo, Govert. 1698. *Letter from G. Bidloo to Antony van Leeuwenhoek about the Animals which are sometimes found in the Liver of Sheep and other Beasts*. Delft.

Bluck, S. & L.J. Levine. 1998. Reminiscence as autobiographical memory: A catalyst for reminiscence theory development. *Aging and Society* 18: 185–208.

Blume, Stuart S. 1992. *Insight and industry: On the dynamics of technological change in medicine. Cambridge*, MA: MIT Press 1992.

Boehm, Gotfried. 1985. *Bildnis und Individuum: über den Ursprung der Porträtmalerei in der Italienischen Reniassance*. München: Prestel.

Boerhaave, Herman. 1741. *Kortbondige Spreuken wegens de Ziektens te Kennen en te Genezen*. Amsterdam.

Bondeson, Jan. 1997. *A cabinet of medical curiosities*. New York: Norton.

——. 2000. *The two-headed boy and other medical marvels*. Ithaca: Cornell University Press.

Bourtchouladze, R. 2002. *Memories are made of this. How memory works in humans and animals*. New York: Columbia University Press.

Bracegirdle, Brian. 1993. The microscopical tradition. In *Companion encyclopedia of the history of medicine*, ed. William F. Bynum and Roy Porter, 102–119. London: Routledge.

Breidbach, Olaf. 1993. Nervenzellen oder Nervennetze? Zur Entstehung des Neuronenkonzepts. In: *Das Gehirn—Organ der Seele? Zur Ideengeschichte der Neurobiologie*, ed. Ernst Florey and Olaf Breidbach, 81–126. Berlin: Akademie.

Breschet, Gilbert, and Augustin R. de Vauzème. 1834. Nouvelles recherches sur la structure de la peau. *Annales des Sciences Naturelles.* Tome second (Septembre, Octobre et Décembre).

——. 1835. *Nouvelles recherches sur la structure de la peau.* Paris: J.-B. Ballière.

Brockhaus. 1898. *Brockhaus' Konversations-Lexicon. Vierzehnte vollständig neubearbeitete Auflage. Revisierte Jubiläums-Ausgabe.* Leipzig: Brockhaus.

Broman, Thomas H. 1996. *The transformation of German academic medicine, 1750–1820.* Cambridge: Cambridge University Press.

Bryson, Norman. 1984. *Tradition and desire: From David to Delacroix.* Cambridge: CUP.

Burri, Regula Valerie. 2008. *Doing images. Zur Praxis medizinischer Bilder.* Bielefeld: Transcript.

Callon, Michel, and Vololona Rabeharisoa. N.d. *Articulating bodies: the case of muscular dystrophies*, http://www.tik.uio.no/Callon_and_Rabeharisoa.pdf.

Cartwright, Lisa. 1995. *Screening the body. Tracing medicine's visual culture.* Minneapolis and London: University of Minnesota Press.

Chadarevian, Soraya de. 1994. Sehen und Aufzeichnen in der Botanik des 19. Jahrhunderts. In *Der Entzug der Bilder. Visuelle Realitäten*, ed. Michael Wetzel and Herta Wolf, 121–44. Munich: Fink.

Chadwick, Edwin. 1877. Skin cleanliness; head to foot washing; public measures proposed for its promotion. *Journal of the Society of Arts* August 17: 883–86.

Chazan, Lilian Krakowski. 2007. *Um meio quilo de gente.* Rio de Janeiro: Fiocruz.

Churchland, P.S. 1986. *Neurophilosophy: toward a unified science of the mind-brain.* Cambridge, MA: MIT Press.

Clark, A. 2001. *Mindware. An introduction to the philosophy of cognitive science.* Oxford: Oxford University Press.

Code, Lorraine. 2004. The power of ignorance. Presented at the *Ethics and Epistemologies of Ignorance* Conference at Pennsylvania State University, March 26–28, 2004.

Coen, Enrico. 1999. *The art of genes. How organisms make themselves.* Oxford: Oxford University Press.

Cohn, Simon, forthcoming. Picturing the brain inside, revealing the illness outside: A comparison of the different meanings attributed to brain scans by scientists and patients. In *Technologized images, technologized bodies: Anthropological approaches to a new politics of vision*, ed. Jeanette Edwards, Penny Harvey and Peter Wade. London: Berghahn.

Coleman, William. 1988. Prussian pedagogy: Purkyne at Breslau, 1823–1839. In *The investigative enterprise: Experimental physiology in nineteenth-century medicine*, ed. William Coleman and Frederic L. Holmes, 15–64. Berkeley California: University of California Press.

Corbin, Alain, Jean Jacques Courtine, and Georges Vigarello, eds. 2005. *Histoire du corps. 2. De la Révolution à la Grand Guerre.* Paris: Seuil.

Crick, Francis. 1994. *Was die Seele wirklich ist. Die naturwissenschaftliche Erforschung des Bewusstseins.* Munich: Artemis.

Csordas, Thomas, ed. 1994. *Embodiment and experience.* Cambridge: Cambridge University Press.

Cunningham, Andrew. 1997 *The Anatomical Renaissance. The resurrection of the anatomical projects of the Ancients.* Aldershot: Ashgate.

——. 2002. The pen and the sword: Recovering the disciplinary identity of physiology and anatomy before 1800. I: Old physiology—the pen. *Studies in History and Philosophy of Biological and Biomedical Sciences* 33: 631–65.

Damasio, A. 1999. *The feeling of what happens. Body and emotion in the making of consciousness.* Orlando: Harcourt.

——. 2003. *Looking for Spinoza. Joy, sorrow, and the feeling brain*. Orlando: Harcourt.
Daston, Lorraine and Peter Galison. 1992. The image of objectivity. *Representations* 40: 81–128.
Daston, Lorraine, ed. 2004. *Things that talk. Object lessons from art and science*. New York: Zone Books.
Davis-Floyd, Robbie. 1992. *Birth as an American rite of passage*. Berkeley: University of California Press.
Dawkins, Richard. 1999. *The selfish gene*. Oxford: Oxford University Press. Originally published in 1976.
Defleur, Melvin L. and Everette E. Dennis. 1998. *Understanding mass communication. A Liberal Arts perspective*. Boston and New York: Houghton Mifflin Company.
Derrida, Jacques. 1987. *The truth in painting*. Trans. G. Bennington & I. McLeod. Chicago: University of Chicago Press. Originally published in 1978.
Descartes, R. 1988. *The Meditations and Selections from the Principles*. La Salle, Illinois: Open Court.
Desjarlais, Robert. 1996. Struggling along. In *Things as they are. New directions in phenomenological anthropology*, ed. Michael Jackson, 70–93. Bloomington: Indiana University Press.
Dijck, J. van. 2004. Mediated memories: Personal cultural memory as an object of cultural analysis. *Continuum. Journal of Media and Cultural Studies* 18: 261–77.
——. 2005. *The transparent body. A cultural analysis of medical imaging*. Seattle and London: Washington University Press.
Dommann, Monika. 2003. *Durchsicht, Einsicht, Vorsicht: Eine Geschichte der Röntgenstrahlen, 1896–1963*. Zurich: Chronos.
Doob, P.R. 1990. *The idea of the labyrinth from Classical Antiquity through the Middle Ages*. Ithaca and London: Cornell University Press.
Draaisma, D. 2000. *Metaphors of memory. A history of ideas about the mind*. Cambridge: Cambridge University Press.
Dreger, Alice Domurat. 2004. *One of us: Conjoined twins and the future of normal*. Cambridge: Harvard University Press.
Duden, Barbara. 1987. *Geschichte unter der Haut. Ein Eisenacher Arzt und seine Patientinnen um 1730*. Stuttgart: Ernst Klett Verlage.
——. 1991. *The woman beneath the skin. A doctor's patients in eighteenth-century Germany*. Cambridge, Mass.: Harvard University Press.
——. 1993. *Disembodying women: perspectives on pregnancy and the unborn*. Cambridge: Harvard University Press.
Duden, Barbara, and Silya Samerski. 1998. Das aufgeschwatzte Risiko—genetische Beratung als Sprach Ritual. *Psychosozial* 21 (Heft 1, nr. 71): 79–88.
Dumit, Joseph. 2004. *Picturing personhood: Brain scans and biomedical identity*. Princeton: Princeton University Press.
Dupras, Joseph. 1996. Brownings 'My last Duchess': Paragon and Parergon. *Papers on Language & Literature* 32, Winter, Issue 1, http://findarticles.com/p/articles/mi_qa3708/is_/ai_n8739275
Eichhorn, Heinrich. 1826. Ueber die Aussonderungen durch die Haut und über die Wege, durch welche sie geschehen. *Archiv für Anatomie und Physiologie*: 405–86.
Eisenberg, Leon. 1977. The search for care. *Daedalus* 106: 235–46.
Fancher, R.E. 1996. *Pioneers of Psychology*. New York—London: W.W. Norton & Company.
Farago, C.J. 1992. *Leonardo da Vinci's "Paragone". A critical interpretation with a new edition of the text in the Codex Urbinas*. Brill's Studies in Intellectual History, Vol. 25. Leiden: Brill.
Feder, Ellen K. 2002. 'Doctor's orders': Parents and intersexed children. In *The subject of care: Feminist perspectives on dependency*, eds. Eva Feder Kittay and Ellen K. Feder, 294–320. Lanham: Rowman & Littlefield Publishers, Inc.

Fend, Mechtild. 2005. Bodily and pictorial surfaces: Skin in French art and medicine, 1790–1860. *Art History* 28(3): 311–39.

Flusser, Vilém. 1992. *Ins Universum der technischen Bilder*. 4th ed. Göttingen: European Photography.

Foerster, Otfried. 1925. Zur Pathogenese und chirurgischen Behandlung der Epilepsie. *Centralblatt für Chirurgie* 25: 531–56.

Foucault, Michel, and Paul Rabinow. 1997. *The essential works of Foucault, 1954–1984*, Vol. 1. New York: The New Press.

Fournier, Marianne. 1991. *The fabric of life: The rise and decline of seventeenth-century microscopy*. Unpublished Ph.D. dissertation: University of Twente.

Franklin, Sarah. 1993/2001. Life Itself: global nature and the genetic imaginary, http://www.comp.lancs.ac.uk/sociology/soc048sf.html.

Freud, Sigmund. 1955. Das Ich und das Es [1923]. In *Gesammelte Werke*. Vol. 13, 3rd ed. London: Imago.

——. 1992. *Zur Auffassung der Aphasien* [1891]. Frankfurt a. M.: Fischer.

Frye, Marilyn. 2004. On ignorant agency: On not knowing what you're doing. Presented at the *Ethics and Epistemologies of Ignorance* Conference at Pennsylvania State University, March 26–28, 2004.

Gammeltoft, Tine. 2007. Sonography and sociality. Obstetric ultrasound imaging in urban Vietnam. *Medical Anthropology Quarterly* 21(2): 133–153.

Gasking, Elizabeth B. 1967. *Investigations into generation 1651–1828*. London: Hutchinson.

Geissler, Paul Wenzel. 2000. The significance of earth-eating: Social and cultural aspects of the geophagy among Luo children. *Africa* 70(4): 653–682.

Geissler, Paul Wenzel, D. Mwaniki, F. Thiong'o, and H. Frijs. 1998. Geophagy, iron status and anaemia among pregnant women on the coast of Kenya. *Transactions of the Royal Society of Tropical Medicine and Hygiene* 92: 549–553.

Georges, Eugenia. 1996. Fetal ultrasound imaging and the production of authoritative knowledge in Greece. *Medical Anthropology Quarterly* 10(2): 157–175.

Gijswijt-Hofstra, Marijke and Willem Frijhoff, eds. 1987. *Nederland betoverd: Toverij en hekserij van de veertiende tot in de twintigste eeuw*. Amsterdam: De Bataafsche Leeuw.

Ginsburg, Faye, and Rayna Rapp, eds. 1995. *Conceiving the New World Order: The global politics of reproduction*. Berkeley: University of California Press.

Gitelman, Lisa. 2006. *Always already new. Media, history and the data of culture*. Cambridge, Mass. and London: MIT Press.

Gomart, Emily, and Antoine Hennion. 1999. A Sociology of attachment: Music amateurs and drug users. In *Actor Network Theory and after*, ed. John Law and J. Hassard, 220-247. Oxford: Blackwell Publishers.

Green, Monica H. 2000. From "Diseases of Women" to "Secrets of Women": The transformation of gynaecological literature in the later Middle Ages. *Journal of Medieval and Early Modern Studies* 30: 5–39.

Grew, Nehemiah. 1684. The description and use of the pores in the skin of the hands and feet, by the learned and ingenious Nehemiah Grew, M.D. Fellow of the College of Physicians and of the Royal Society. *Philosophical Transactions* 14: 566–67.

Grunenberg, Christoph, and Victoria Pomery. 2002. *Marc Quinn*. Liverpool: Tate Liverpool.

Guillaumin, Colette. 1995. Race and nature: The system of marks. *Racism, sexism, power and ideology*, 133–152. London: Routledge Press.

Gumbrecht, Hans Ulrich. 2004. *Productions of presence. What meaning cannot convey*. Stanford: Stanford University Press.

Gurlt, Ernst F. 1835. Vergleichende Untersuchungen über die Haut des Menschen und der Haus-Säugethiere, besonders in Beziehung auf die Absonderungs-Organe des Haut-Talges und des Schweisses. *Archiv für Anatomie, Physiologie und wissenschaftliche Medicin*: 399–418.

Hagner, Michael. 1994. Lokalisation, Funktion, Cytoarchitektonik. Wege zur Modellierung des Gehirns. In *Objekte, Differenzen und Konjunkturen*, ed. Michael Hagner, Hans-Jörg Rheinberger and Bettina Wahrig-Schmidt, 121–50. Berlin: Akademie.

——. 2002. Cyber-Phrenologie. Die neue Physiognomik des Geistes und ihre Ursprünge. In *Die Politik der Maschine*, ed. Klaus Peter Dencker, 182–98. Hamburg: Verlag Hans-Brednow-Institut.

——. 2003. Scientific medicine. In *From natural philosophy to the sciences*, ed. David Cahan, 49–87. Chicago, London: The University of Chicago Press.

——. 2004. *Geniale Gehirne. Zur Geschichte der Elitegehirnforschung*. Göttingen: Wallstein.

——. 2006. *Der Geist bei der Arbeit. Historische Untersuchungen zur Hirnforschung*. Göttingen: Wallstein.

Hansen, Julie. 1996(a). Resurrecting death: Anatomical art in the Cabinet of Dr. Frederik Ruysch. *Art Bulletin* 78: 663–679.

——. 1996(b). *Galleries of life and death: The anatomy lesson in Dutch art, 1603–1777* (Unpublished Ph.D. dissertation: Stanford University).

Hansen, Mark B.N. 2006. *Bodies in code. Interfaces with digital media*. New York and London: Routledge.

Haraway, Donna. 1997. *Modest_Witness@second_Millenium.FemaleMan©_Meets_Oncomouse™. Feminism and Technoscience*. London & New York: Routledge.

Harding, Sandra. 2004. Two influential theories of ignorance and philosophy's interests in ignoring them. Presented at the *Ethics and Epistemologies of Ignorance* Conference at Pennsylvania State University, March 26–28, 2004.

Hayles, Katherine. 1999. *How we became posthuman. Virtual bodies in cybernetics, literature, and informatics*. Chicago & London: UCP.

Heidegger, Martin. 1996. *Being and Time*. Trans. Joan Stambaugh. Albany: SUNY Press.

Heijden, Cor van der. 2001. *Rampen en plagen in Nederland 1400–1940. Pestbacillen, paalwormen en plunderende Moskovieters*. Zaltbommel: Kempen.

Helen Chadwick. Enfleshings. 1989. With an essay from M. Warner. London: Seckler & Warburg.

Helman, Cecil G. 2001. *Culture, health and illness*. Fourth edition (first edition 1984). London: Arnold.

Henle, Friedrich G.J. 1841. *Allgemeine Anatomie-Lehre von den Mischungs—und Formbestandtheilen des menschlichen Körpers*. Leipzig: Leopold Voß.

Hoffmeyer, Jesper. 2006. Genes, development, and semiosis. In *Genes in Development. Re-reading the molecular paradigm*, ed. Eva Neumann-Held and Christoph Rehmann-Sutter, 152–175. Durham: Duke University Press.

Houston News. 2003. Twin Boys Begin New Lives After Successful Separation. http://www.click2houston.com/news/2550089/detail.html.

Hüsing, Bärbel, Lutz Jäncke, and Brigitte Tag. 2006. *Impact assessment of neuroimaging*. Zürich: IVT Verlag.

Husserl, Edmund. 1982. *Ideas pertaining to a pure phenomenology and to a phenomenological philosophy*. First book: General introduction to a pure phenomenology. Trans. F. Kersten and André Schuwer. Dordrecht: Kluwer Academic Publishers.

——. 1989. *Ideas pertaining to a pure phenomenology and to a phenomenological philosophy*. Second book: Studies in the phenomenology of constitution. Trans. Richard Rojcewicz and André Schuwer. Dordrecht: Kluwer Academic Publishers.

Hutchins, E. 1996. *Cognition in the Wild*. Cambridge: MIT Press.

Hydén, Lars-Christer. 1997. Illness and narratives. *Sociology of health and illness* 19 (1): 48–69.

Ihde, Don. 1990. *Technology and the life world*, Bloomington and Minneapolis: Indiana University Press.

——. 2002. *Bodies in technology*, Minneapolis and London: Minnesota University Press.

Intersex Society of North America (ISNA) website: www.isna.org.

Israel, Jonathan. 1995. *The Dutch Republic. Its rise, greatness and fall 1477–1806*. Oxford: Clarendon.

Johnson, Barbara. 1980. "The Frame of reference: Poe, Lacan, Derrida". In *The critical difference: Essays in the contemporary rhetoric of reading*, 110–146. Baltimore: John Hopkins UP.

Johnson, S. 2004a. *Mind wide open: Your brain and the neuroscience of everyday life*. New York: Scribner.

——. 2004b. The science of Eternal Sunshine. You can't erase your boyfriend from your brain, but the movie gets the rest of it right. In: Slate.msn.com, posted March 22, 2004. http://slate.msn.com/id/2097502/. (accessed April 18, 2006).

Jones, Jonathan. 2001. John Sulston, Marc Quinn, *The Guardian* 22.09.2001. no pages, http://arts.guardian.co.uk/portrait/story/o,,7440345,00.html#article_continue

Jütte, Robert. 1996. Die Frau, die Kröte und der Spitalmeister. Zur Bedeutung der etnographischen Methode für eine Sozial- und Kulturgeschichte der Medizin. *Historische Anthropologie* 4: 193–215.

Kay, Lily. 2000. *Who wrote the Book of Life? A history of the genetic code*. Stanford: SUP.

Kevles, Daniel, and Leroy Hood, eds. 1990. *The Code of Codes: Scientific and social issues in the Human Genome Project*. Cambridge Ma.: Harvard University Press.

Kleinman, Arthur. 1980. *Patients and healers in the context of culture*. Berkeley: University of California Press.

——. 1988. *The illness narratives. Suffering, healing, and the human condition*. New York: Basic Books.

Knoeff, Rina. 2006. Chemistry, mechanics and the making of anatomical knowledge. Boerhaave vs. Ruysch on the nature of the glands. *Ambix* 53: 201–219.

Koch, Gertrud. 1985. Ex-Changing the Gaze: Re-Visioning Feminist Film Theory. *New German Critique*, Winter 1985: 139–153.

Kooijmans, Luuc. 2004. *De doodskunstenaar. De anatomische lessen van Frederik Ruysch*. Amsterdam: Bert Bakker.

Kölliker, Albert von. 1850. *Mikroskopische Anatomie oder Gewebelehre des Menschen 2.Band 1.Hälfte: Specielle Gewebelhre von der Haut, den Muskeln, Knochen und Nerven*. Leipzig: Wilhelm Engelmann Verlag.

Krause, Carl F.T. 1844. Haut. In *Handwörterbuch der Physiologie: mit Rücksicht auf physiologische Pathologie*, ed. Rudolph Wagner, Zweiter Band: 108–86. Braunschweig: Vieweg.

Kruta, Vladislav. 1975. Purkyne (Purkinje), Jan Evangelista. In *Dictionary of scientific biography*, ed. Charles C. Gillisepie, 214. New York: Charles Scribner's Sons.

Kuriyama, Shigehisa. 1999. *The expressiveness of the body and the divergence of Greek and Chinese medicine*. New York: Zone Books.

La Berge, Anne. 1994. Medical microscopy in Paris, 1830–1855. In *French medical culture in the nineteenth century*, ed. Anne La Berge and Mordechai Feingold, 296–326. Amsterdam, Atlanta G.A.: Rodopi.

——. 1998. Dichotomy or integration? Medical microscopy and the Paris clinical tradition. In *Constructing Paris medicine*, ed. Caroline Hannaway and Anne La Berge, 275–312. Amsterdam, Atlanta G.A.: Rodopi.

——. 1999. The history of science and the history of microscopy. *Perspectives on Science* 7(1): 111–42.

Latour, Bruno. 1999. *Pandora's hope. Essays on the reality of science studies*. Cambridge, Mass. and London: Harvard University Press.

——. 2004. How to talk about the body? The normative dimension of science studies. *Body & Society* 10 (2–3): 205–229.

——. 2005. *Re-assembling the social: an introduction to actor-network-theory*. Oxford: Oxford University Press.

Lawrence, Christopher and Steven Shapin, eds. 1998. *Science incarnate. Historical embodiments of natural knowledge*. Chicago: Chicago University Press.

Leder, Drew. 1999. *The absent body*. Chicago and London: University of Chicago Press.
Leeuwenhoek, Antoni van. 1693. An extract of a letter from Mr. Anthony van Leeuwenhoek, to the R.S. containing his observations on the seeds of cotton, palm, or date-stones, cloves, nutmegs, goose-berries, currans, tulips, cassia, lime-tree: On the skin of the hand, and pores, of sweat, the crystalline humour, optic nerves, gall, and scales of fish: and the figures of several salt particles, etc. *Philosophical Transactions* 17: 949–60.
Leroi, Armand. 2003. *Mutants. On the form, varieties and errors of the human body*. New York: Viking.
Leslie, E. 1999. Souvenirs and Forgetting: Walter Benjamin's Memory-work. In *Material Memories*, ed. M. Kwint, C. Breward and J. Aynsley, 107–23. Oxford: Berg.
Levine, L.J. 1997. Reconstructing memories for emotions. *Journal of Experimental Psychology* 126: 176–7.
Linnaeus, Carl. 1995 (1732). *The Lapland Journey*. Trans. Peter Graves. Edinburgh: Lockharton Press.
Maclachlan, Malcolm. 2004. *Embodiment. Clinical, critical and cultural perspectives*. Maidenhead: Open University Press.
Mann, Th. [1924] 1987. *Der Zauberberg*. Berlin: Fischer Verlag.
Mannix, Daniel P. 1999. *Freaks: We who are not as others*. New York: Re/Search Publications, Inc.
Martell, Joanne. 2000. *Millie-Christine: Fearfully and wonderfully made*. Winston-Salem: John F. Blair, Publisher.
Martin, L.H. 1988. *Technologies of the self: A seminar with Michel Foucault*. London: Tavistock.
Merleau-Ponty, M. 1962. *Phenomenology of perception*. Trans. Colin Smith. London & New York: Routledge & Kegan Paul.
——. 1968. *The visible and the invisible*. Evanston: Northwestern University Press.
Mitchell, Lisa. 2001. *Baby's first picture: Ultrasound and the politics of fetal subjects*. Toronto: University of Toronto Press.
Mohl, Hugo von. 1846. *Mikrographie oder Anleitung zur Kenntniss und zum Gebrauche des Mikroskops*. Tübingen: L.F. Fues.
Mol, Annemarie. 2002. *The body multiple. Ontology in medical practice*. Durham and London: Duke University Press.
Mol, Annemarie and Law, John. 2004. Embodied action, enacted bodies: The example of hypoglycaemia. *Body & Society* 10(2–3): 43–62.
Monakow, Constantin von. 1897. *Gehirnpathologie*. Wien: Hölder.
Müller-Rockstroh, Babette. 2007. *Ultrasound travels. The politics of a medical technology in Ghana and Tanzania*. PhD diss., Maastricht University.
Munk, Hermann. 1890. *Über die Functionen der Grosshirnrinde. Gesammelte Mitteilungen*. 2nd ed. Berlin: Hirschwald.
Nancy, J.-L. 1979. *Ego Sum*. Paris: Flammarion.
——. 1996. *Être singulier pluriel*. Paris: Galilée.
——. 2000. *L'intrus*. Paris, Galilée.
——. 2006a. De l'âme. In *Corpus*. 107–129, Paris: Édition Métaillié.
——. 2006a. L'extension de l'âme. In *Corpus*. 130–144, Paris: Édition Métaillié.
NBC 4. 2003. Conjoined Twins Remain in Serious Condition After Separation Surgery. http://www.nbc4.tv/print/2475058/detail.html.
Netta, Irene. 2005. "Between Representation and Abstraction". In *Andreas Horlitz. Arbeiten Works*, 104–146. Nürnberg: Verlag für Moderne Kunst.
Neumann, Isidor. 1869. *Lehrbuch der Hautkrankheiten*. Wien: Wilhelm Braumüller.
Neumann-Held, Eva and Christoph Rehmann-Sutter, eds. 2006. *Genes in development. Re-reading the molecular paradigm*. Durham: Duke University Press.
Outram, Dorinda. 1989. *The body and the French Revolution: Sex, class and political culture*. London: Yale University Press.

Park, Katherine. 2006. *Secrets of women Gender, generation, and the origins of human dissection*, New York: Zone Books.

Pater, Walter. 1980. *The Renaissance. Studies in art and poetry*. The 1893 text. Edited, with textual and explanatory notes by Donald I. Hill, Berkeley, Los Angeles, London: University of California Press.

Penfield, Wilder and Theodore Rasmussen. 1950. *The cerebral cortex of man. A clinical study of localization of function*. New York: Macmillan.

Petchesky, Rosalind. 1987. Fetal images: The ower of visual culture in the politics of reproduction. *Feminist Studies* 2(2): 263–293.

Pickstone, John V. 2000. *Ways of knowing: A new history of science, technology and medicine*. Chicago: University of Chicago Press.

Pilloud, Séverine, and Micheline Louis-Courvoisier. 2003. The intimate experience of the body in the eighteenth century: Between interiority and exteriority. *Medical History* 47(4): 451–72.

Pomian, Krzysztof. 1997. *Der Ursprung des Museums. Vom Sammeln*. 3rd ed. Berlin: Wagenbach.

Porter, Roy. 1985. The patient's view: Doing medical history from below. *Theory and Society*, 14: 175–98.

Posner, Michael I., and Marcus E. Raichle. 1994. *Images of mind*. New York: Scientific American Library.

Prentice, Rachel. 2005. The anatomy of a surgical simulation: The mutual articulation of bodies in and through the machine. *Social Studies of Science* 35(6): 837–66.

Price, J.L. 1998. *The Dutch Republic in the seventeenth century*. New York: St. Martin's Press.

Purkyne, Jan E. 1823. *Commentatio de Examine Physiologico Organi Visus et Systematis Cutanei*. Vratislaviae: Typis Universitatis.

——. 1844. Mikroskop. In *Handwörterbuch der Physiologie: mit Rücksicht auf physiologische Pathologie*, ed. Rudolph Wagner, Zweiter Band, 411–41. Braunschweig: Vieweg.

Purkyne, Jan E., Vladislav Kruta, Joachim Ebert, and Karel Zlábek. 1979. *Abhandlung über die physiologische Untersuchung des Sehorgans und des Hautsystems / von Johann Evangelista Purkyně; übers. von Joachim Ebert und Karel Žlábek; eingeleitet und erläutert von Vladislav Kruta*. Halle, Saale: Deutsche Akademie der Naturforscher Leopoldina.

Radstake, Maud. 2007. *Visions of illness. An endography of real-time medical imaging*. Delft: Eburon.

Rheinberger, Hans-Jörg. 1997. "Von der Zelle zum Gen. Repräsentationen der Molecularbiologie". In *Räume des Wissens. Repräsentation, Codierung, Spur*, ed. Hans-Jörg Rheinberger, Michael Hagner and Bettina Wahrig-Schmidt, 265–281. Berlin: Akademie Verlag.

——. 2005. *Iterationen*. Berlin: Merve Verlag.

Richter, J.P. 1883. *The literary works of Leonardo da Vinci*, London: Sampson Low, Marston, Searle & Rivington.

Ricoeur, P. 1991. Narrative Identity. In *On Paul Ricoeur. Narrative and Interpretation*, ed. D. Wood, 188–199. London and New York: Routledge.

Rosen, Philip, ed., 1986. *Narrative, apparatus, ideology. A film theory reader*. New York: Columbia University Press.

Roth, Gerhard. 1994. Braucht die Hirnforschung die Philosophie? In *Neuroworlds. Gehirn—Geist—Kultur*, ed. Jutta Fedrowitz, Dirk Matejovski, and Gert Kaiser, 81–92. Frankfurt a. M.: Campus.

Rothman, Barbara Katz. 1986. *Tentative pregnancy: How amniocentesis changes the experience of motherhood*. New York: Norton.

Ruestow, Edward G. 1996. *The microscope in the Dutch Republic: The shaping of discovery*. Cambridge: Cambridge University Press.

Ruysch, Frederik. 1744. *Alle de Genees—en Heelkundige Werken van Frederik Ruysch*, 3 vols. Amsterdam.

Ryle, G. 1949. *The concept of mind*. Harmondsworth, Penguin University Books.

Saetnan, Ann, Nelly Oudshoorn, and Martha Kirejczyk, eds. 2000. *Bodies of technology. Women's involvement with reproductive medicine*. Columbus, Ohio: Ohio State University Press.

Sarasin, Philipp. 2001. *Reizbare Maschinen: eine Geschichte des Körpers 1765–1914*. Frankfurt am Main: Suhrkamp.

Sawday, Jonathan. 1995. *The body emblazoned: Dissection and the human body in Renaissance culture*. London: Routledge.

Schama, Simon. 1991. *The embarrassment of riches. An interpretation of Dutch culture in the Golden Age*. London: Fontana.

Scheper-Hughes, Nancy, and Margaret Lock. 1987. The mindful body: A prolegomenon to future work in medical anthropology. *Medical Anthropology Quarterly* 1(1): 6–41.

Schickore, Jutta. 2001. Eröffnung der Augen: Auge und Sehen in der mikroskopischen Anatomie. In *Wahrnehmung der Natur-Natur der Wahrnehmung: Studien zur Geschichte visueller Kultur um 1800*, ed. Gabriele Dürbeck, Jutta Schickore and Gerhard Wiesenfeldt, 165–77. Dresden: Verlag der Kunst.

——. 2003. The 'philosophical grasp of the appearances' and experimental microscopy: Johannes Müller's microscopic research, 1824–1832. *Studies in History and Philosophy of Biological and Biomedical Sciences* 34: 569–92.

——. 2007. *The microscope and the eye: A history of reflections, 1740–1870*. Chicago: the University of Chicago Press.

Schott, G.D. 1993. Penfield's homunculus: a note on cerebral cartography. *Journal of Neurology, Neurosurgery and Psychiatry* 56: 329–333.

Scully, Jackie Leach. 2006. Nothing Like a Gene. In *Genes in development. Re-reading the molecular paradigm*, ed. E. Neumann-Held and C. Rehmann-Sutter, 349–365, Durham: Duke University Press.

Sedgwick, Eve Kosofsky. 1990. *Epistemology of the closet*. Berkeley: University of California Press.

Sherwood-Smith, Maria. 1997. God and gynaecology: Women's secrets in the Dutch "Historiebijbel van 1360". *German Life and Letters* 50: 390–402.

——. 1999. Forschung oder Vorurteil, Kultur oder Naturkunde. Zur Frage der Frauenfeindlichkeit in den deutschen und niederländischen Bearbeitungen der "Secreta Mulierum" von Pseudo Magnus. Natur und Kultur in der deutschen Literatur des Mittelalters. In *Natur und Kultur in der deutschen Literatur des Mittlealters*, ed. Alan Robertshaw and Gerhard Wolf, Gerhard, 163–174. (Tübingen).

Silverman, Kaja. 1992. *Male subjectivity at the margins*. New York: Routledge.

Simon, Gustav. 1848. *Die Hautkrankheiten durch anatomische Untersuchungen erläutert*. Berlin: G. Reimer.

Slater, D. 1995. Domestic photography and digital culture. In: *The photographic image in digital culture*, ed. M. Lister, 129–46. New York: Routledge.

Slatman, Jenny. 2005a. De zichtbaarheid van het denken. Over de visualisering van hersenactiviteit in wetenschap en kunst. *Feit & Fictie. Tijdschrift voor de geschiedenis van de representatie* VI (2): 108–115.

——. 2005b. The sense of life: Husserl and Merleau-Ponty on touching and being touched. *Chiasmi International*, 7: 305–325.

——. 2007. Recognition beyond narcissism. Imaging the body's ownness and strangeness. In *The Other. Feminist Reflections in Ethics*, ed. H. Fielding, G. Hiltmann, D. Olkowski and A. Reichold, 186–204. London: Palgrave.

——. 2008. *Vreemd lichaam. Over medisch ingrijpen en persoonlijke identiteit*. Amsterdam: Ambo.

Smith, Pamela. 2004. *The body of the artisan. Art and experience in the Scientific Revolution*. Chicago: University of Chicago Press.

Sobchack, Vivian. 2004. The scene of the screen: Envisioning photographic, cinematic and electronic 'presence'. In: *Carnal thoughts. Embodiment and moving image culture*, 135–162. Berkeley, Los Angeles and London: University of California Press.

Soemmerring, Samuel Thomas. 1791. Abbildungen und Beschreibungen einiger Misgeburten. Mainz: Universitätsbuchhandlung Stafford, Barbara M. 1991. *Body criticism. Imaging the unseen in Enlightenment art and medicine*. Cambridge Mass.: MIT.

Sontag, Susan. 1991. *Illness as metaphor and AIDS and its metaphors*. London: Penguin Books.

Stabile, Carol. 1992. Shooting the mother: Fetal photography and the politics of disappearance. *Camera Obscura* 28: 180–205.

Stefaniak, Regina. 1997. On looking into the abyss: Leonardo's *Virgin of the Rocks*, *Konsthistorisk Tidskrift*, 66: 1–36.

Stolberg, Michael. 2000. An unmanly vice. Self-pollution, anxiety and the body in the eighteenth century. *Social History of Medicine* 13: 1–21.

——. 2003. *Homo Patiens. Krankheits- und Körpererfahrung in der Frühen Neuzeit*. Köln, Weimar, Vienna: Böhlau Verlag.

Strathern, Marilyn. (forthcoming). Can one rely on knowledge? In *Ethnography of medical research in Africa*, ed. Wenzel Geissler and Catherine Molyneux. London: Berghahn.

Strauss, Darin. 2000. *Chang and Eng: A novel*. New York: Dutton.

Sturken, Marita and Lisa Cartwright. 2001. *Practices of looking. An introduction to visual culture*. Oxford: Oxford University.

Sutton, J. 1998. *Philosophy and memory traces. Descartes to connectionism* Cambridge: Cambridge University Press.

——. 2002. Porous Memory and the Cognitive Life of Things. In *Prefiguring cyberculture. An intellectual history*, ed. D. Tofts, A. Jonson, and A. Cavallara. 130–41. Cambridge: MIT Press.

Swan, Claudia. 2005. *Art, science, and witchcraft in early modern Holland. Jaques de Gheijn II (1565–1629)*. Cambridge: Cambridge University Press.

Tautz, Sigrid, Albrecht Jahn, Ibrahim Molokomme, and Regina Joergen. 2000. Between fear and relief: How pregnant women experience foetal ultrasound in a Botswana District Hospital. *Social Science and Medicine* 50(5): 689–701.

Taylor, Janelle. 1992. The public fetus and the family car. From abortion politics to a Volvo advertisement. *Public Culture* 4(2): 67–80.

Taylor, John. 1994. Windows open on the mind over grey matter. *The Times Higher Education Supplement, Multimedia* December 9: iv–v.

Theopold, Wilhelm. 1984. *Mirakel. Heilung zwischen Wissenschaft und Glauben*. Munich.

Thomson, Rosemarie Garland. 1997. *Extraordinary bodies: Figuring physical disability in American culture and literature*. New York: Columbia University Press.

Toombs, S. Kay. 1992. *The meaning of illness. A phenomenological account of the different perspectives of physician and patient*. Dordrecht: Kluwer Academic Publishers.

Tuana, Nancy. 2004. Coming to Understand: Orgasm and the Epistemology of Ignorance. *Hypatia: A Journal of Feminist Philosophy* vol. 19, no. 1 (Winter 2004): 194–232.

Verbeek, Peter Paul. 2005. *What things do: Philosophical reflections on technology, agency, and design*, trans. R.P. Crease, University Park, PA: Pennsylvania State University Press.

Vogel, Julius. 1841. *Anleitung zum Gebrauch des Mikroskopes zur zoochemischen Analyse und zur mikroskopisch-chemische Untersuchung überhaupt*. Leipzig: Leopold Voss.

Wagner, R. 1839. *Erläuterungstafeln zur Physiologie und Entwickelungsgeschichte*. Leipzig: Leopold Voss.

Weber, Ernst H. 1827. Beobachtungen über die Oberhaut, die Hautbälge und ihre Vergrösserung in Krebsgeschwülsten und über die Haare des Menschen. *Archiv für Anatomie und Physiologie*: 198–225.

Weigel, Sigrid. 2004. Phantombilder zwischen Messen und Deuten. Bilder von Hirn und Gesicht in den Instrumentarien empirischer Forschung von Psychologie und Neurowissenschaft. In *Repräsentationen. Medizin und Ethik in Literatur und Kunst der Moderne*, ed. Bettina von Jagow and Florian Steger, 159–99. Heidelberg: Winter.

Weiss, Gail. 1999. *Body images: Embodiment as intercorporeality*, New York and London: Routledge.

Wendt, Alphons. 1834. Ueber die menschliche Epidermis. *Archiv für Anatomie, Physiologie und wissenschaftliche Medicin*: 278–91.

Wernicke, Carl. 1874. *Der aphasische Symptomencomplex. Eine psychologische Studie auf anatomischer Basis*. Breslau: Cohn & Weigert.

Wilson, Catharine. 1995. *The invisible world: Early modern philosophy and the invention of the microscope*. Princeton: Princeton University Press.

Winkler, H. 2002. Discourses, schemata, technology, monuments: Outline for a theory of cultural continuity. *Configurations*, 10: 91–109.

Wittgenstein, L. 1953. *Philosophical investigations*. Oxford: Blackwell.

Working drafts. Envisioning the human genome. 2002. London: The Wellcome Trust.

Young, Iris Marion. 1984. Pregnant embodiment: Subjectivity and alienation. *The Journal of Medicine and Philosophy* 9: 45–62.

Young, Katherine. 1989. Disembodiment: the phenomenology of the body in medical examinations. *Semiotica* 73(1/2): 43–66.

Zedler, Johann H. 1732–1754. *Grosses vollständiges Universal-Lexicon Aller Wissenschaften und Künste*. Leibzig.

Zwijnenberg, Robert. 1999. *The writings and drawings of Leonardo da Vinci. Order and chaos in early modern thought*. Cambridge University Press.

INDEX

Page numbers in bold refer to illustrations.

Brill's Studies in Intellectual History

Series Editor: A.J. Vanderjagt

161. Stern-Gillet, S. & K. Corrigan (eds.). *Reading Ancient Texts. Volume I: Presocratics and Plato*. Essays in Honour of Denis O'Brien. 2007. ISBN 978 90 04 16509 0
162. Stern-Gillet, S. & K. Corrigan (eds.). *Reading Ancient Texts. Volume II: Aristotle and Neoplatonism*. Essays in Honour of Denis O'Brien. 2007. ISBN 978 90 04 16512 0
164. Stefaniak, R. '*Mysterium Magnum*'. Michelangelo's Tondo Doni. 2008. ISBN 978 90 04 16544 1 (Published as Vol. 1 in the subseries *Brill's Studies on Art, Art History, and Intellectual History*)
165. Catana, L. *The Historiographical Concept 'System of Philosophy'*. Its Origin, Nature, Influence and Legitimacy. 2008. ISBN 978 90 04 16648 6
166. Goodare, J. & A.A. MacDonald (eds.). *Sixteenth-Century Scotland*. Essays in Honour of Michael Lynch. 2008. ISBN 978 90 04 16825 1
167. Van Bunge, W. & H. Bots (eds.). *Pierre Bayle (1647-1706)*, le philosophe de Rotterdam: *Philosophy, Religion and Reception*. Selected Papers of the Tercentenary Conference held at Rotterdam, 7-8 December 2006. 2008. ISBN 978 90 04 16536 6
168. Dixhoorn, A. van & S. Speakman Sutch. *The Reach of the Republic of Letters*. Literary and Learned Societies in Late Medieval and Early Modern Europe (2 Vols.). 2008. ISBN 978 90 04 16955 5
169. Cohen, S. *Animals as Disguised Symbols in Renaissance Art*. 2008. ISBN 978 90 04 17101 5 (Published as Vol. 2 in the subseries *Brill's Studies on Art, Art History, and Intellectual History*)
170. Lennon, T.M. *The Plain Truth*. Descartes, Huet, and Skepticism. 2008. ISBN 978 90 04 17115 2
171. Lehner, U.L. (ed.). Beda Mayr, *Vertheidigung der katholischen Religion*. Sammt einem Anhange von der Möglichkeit einer Vereinigung zwischen unserer, und der evangelisch-lutherischen Kirche (1789). 2009. ISBN 978 90 04 17318 7 (Published as Vol. 5 in the subseries *Brill's Texts and Sources in Intellectual History*)
172. Rothkamm, J. *Institutio Oratoria*. Bacon, Descartes, Hobbes, Spinoza. 2009. ISBN 978 90 04 17328 6
173. Richardson, C.M. *Reclaiming Rome*. Cardinals in the Fifteenth Century. 2009. ISBN 978 90 04 17183 1
174. Dyson, R.W. (ed.). *James of Viterbo*: De regimine Christiano. A Critical Edition and Translation. 2009. ISBN 978 90 04 17597 6 (Published as Vol. 6 in the subseries *Brill's Texts and Sources in Intellectual History*)
175. Lærke, M. (ed.). *The Use of Censorship in the Enlightenment*. 2009. ISBN 978 90 04 17558 7
176. Vall, R. van de & R. Zwijnenberg (eds.). *The Body Within*. Art, Medicine and Visualization. 2009. ISBN 978 90 04 17621 8 (Published as Vol. 3 in the subseries *Brill's Studies on Art, Art History, and Intellectual History*)
177. Sauter, M.J. *Visions of the Enlightenment*. The Edict on Religion of 1788 and the Politics of the Public Sphere in Eighteenth-Century Prussia. 2009. ISBN 978 90 04 17651 5
178. Verbaal, W., Y. Maes & J. Papy (eds.). *Latinitas Perennis*. Volume II: Appropriation by Latin Literature. 2009. ISBN 978 90 04 17683 6
179. Miert, D.K.W. van. *Humanism in an Age of Science*. The Amsterdam Athenaeum in the Golden Age, 1632-1704. 2009. ISBN 978 90 04 17685 0
180. Pietarinen J. & V. Viljanen (eds.). *The World as Active Power*. Studies in the History of European Reason. 2009. ISBN 978 90 04 17712 3.
181. Maia Neto, J.R., J.C. Laursen & G. Paganini (eds.). *Skepticism in the Modern Age*. Building on the Work of Richard Popkin. 2009. ISBN 978 90 04 17784 0
182. Cruz, L. & W. Frijhoff (eds.). *Myth in History, History in Myth*. 2009. ISBN 978 90 04 17834 2

www.ingramcontent.com/pod-product-compliance
Lightning Source LLC
LaVergne TN
LVHW080847300926
849162LV00003B/364